COMPUTER-AIDED DESIGN TECHNIQUES FOR LOW POWER SEQUENTIAL LOGIC CIRCUITS

THE KLUWER INTERNATIONAL SERIES IN ENGINEERING AND COMPUTER SCIENCE

VLSI, COMPUTER ARCHITECTURE AND DIGITAL SIGNAL PROCESSING
Consulting Editor
Jonathan Allen

Other books in the series:

APPLICATION SPECIFIC PROCESSORS
  E.E. Swartzlander, Jr.
  ISBN: 0-7923-9729
QUICK-TURNAROUND ASIC DESIGN IN VHDL: Core-Based Behavioral Synthesis
  M.S. Romdhane, V.K. Madisetti, J.W. Hines
  ISBN: 0-7923-9744-4
ADVANCED CONCEPTS IN ADAPTIVE SIGNAL PROCESSING
  W. Kenneth Jenkins, Andrew W. Hull, Jeffrey C. Strait
  ISBN: 0-7923-9740-1
SOFTWARE SYNTHESIS FROM DATAFLOW GRAPHS
  Shuvra S. Bhattacharyya, Praveen K. Murthy, Edward A. Lee
  ISBN: 0-7923-9722-3
AUTOMATIC SPEECH AND SPEAKER RECOGNITION: Advanced Topics,
  Chin-Hui Lee, Kuldip K. Paliwal
  ISBN: 0-7923-9706-1
BINARY DECISION DIAGRAMS AND APPLICATIONS FOR VLSI CAD, Shin-ichi Minato
  ISBN: 0-7923-9652-9
ROBUSTNESS IN AUTOMATIC SPEECH RECOGNITION, Jean-Claude Junqua, Jean-Paul Haton
  ISBN: 0-7923-9646-4
HIGH-PERFORMANCE DIGITAL VLSI CIRCUIT DESIGN, Richard X. Gu, Khaled M. Sharaf, Mohamed I. Elmasry
  ISBN: 0-7923-9641-3
LOW POWER DESIGN METHODOLOGIES, Jan M. Rabaey, Massoud Pedram
  ISBN: 0-7923-9630-8
MODERN METHODS OF SPEECH PROCESSING, Ravi P. Ramachandran
  ISBN: 0-7923-9607-3
LOGIC SYNTHESIS FOR FIELD-PROGRAMMABLE GATE ARRAYS, Rajeev Murgai, Robert K. Brayton
  ISBN: 0-7923-9596-4
CODE GENERATION FOR EMBEDDED PROCESSORS, P. Marwedel, G. Goossens
  ISBN: 0-7923-9577-8
DIGITAL TIMING MACROMODELING FOR VLSI DESIGN VERIFICATION, Jeong-Taek Kong, David Overhauser
  ISBN: 0-7923-9580-8
DIGIT-SERIAL COMPUTATION, Richard Hartley, Keshab K. Parhi
  ISBN: 0-7923-9573-5
FORMAL SEMANTICS FOR VHDL, Carlos Delgado Kloos, Peter T. Breuer
  ISBN: 0-7923-9552-2

COMPUTER-AIDED DESIGN TECHNIQUES FOR LOW POWER SEQUENTIAL LOGIC CIRCUITS

by

José Monteiro
Massachusetts Institute of Technology

and

Srinivas Devadas
Massachusetts Institute of Technology

KLUWER ACADEMIC PUBLISHERS
Boston / Dordrecht / London

Distributors for North America:
Kluwer Academic Publishers
101 Philip Drive
Assinippi Park
Norwell, Massachusetts 02061 USA

Distributors for all other countries:
Kluwer Academic Publishers Group
Distribution Centre
Post Office Box 322
3300 AH Dordrecht, THE NETHERLANDS

Library of Congress Cataloging-in-Publication Data

A C.I.P. Catalogue record for this book is available
from the Library of Congress.

Copyright © 1997 by Kluwer Academic Publishers

All rights reserved. No part of this publication may be reproduced, stored in a retrieval system or transmitted in any form or by any means, mechanical, photo-copying, recording, or otherwise, without the prior written permission of the publisher, Kluwer Academic Publishers, 101 Philip Drive, Assinippi Park, Norwell, Massachusetts 02061

Printed on acid-free paper.

Printed in the United States of America

Contents

Table of Contents ... v

List of Figures ... ix

List of Tables ... xiii

Preface ... xv

Acknowledgments ... xvii

1 Introduction ... 1
  1.1 Power as a Design Constraint ... 3
  1.2 Organization of this Book ... 5
       References ... 6

2 Power Estimation ... 9
  2.1 Power Dissipation Model ... 10
  2.2 Switching Activity Estimation ... 11
       2.2.1 Simulation-Based Techniques ... 11
       2.2.2 Issues in Probabilistic Estimation Techniques ... 13
       2.2.3 Probabilistic Techniques ... 16
  2.3 Summary ... 19
       References ... 20

3 A Power Estimation Method for Combinational Circuits ... 23
  3.1 Symbolic Simulation ... 24
  3.2 Transparent Latches ... 27
  3.3 Modeling Inertial Delay ... 29

	3.4	Power Estimation Results	30
	3.5	Summary	32
		References	33

4 Power Estimation for Sequential Circuits — 35
- 4.1 Pipelines . . . 35
- 4.2 Finite State Machines: Exact Method . . . 37
  - 4.2.1 Modeling Temporal Correlation . . . 38
  - 4.2.2 State Probability Computation . . . 39
  - 4.2.3 Power Estimation given State Probabilities . . . 41
- 4.3 Finite State Machines: Approximate Method . . . 42
  - 4.3.1 Basis for the Approximation . . . 42
  - 4.3.2 Computing Present State Line Probabilities . . . 43
  - 4.3.3 Picard-Peano Method . . . 45
  - 4.3.4 Newton-Raphson Method . . . 47
  - 4.3.5 Improving Accuracy using m-Expanded Networks . . 52
  - 4.3.6 Improving Accuracy using k-Unrolled Networks . . . 53
  - 4.3.7 Redundant State Lines . . . 54
- 4.4 Results on Sequential Power Estimation Techniques . . . 57
- 4.5 Modeling Correlation of Input Sequences . . . 65
  - 4.5.1 Completely and Incompletely Specified Input Sequences 66
  - 4.5.2 Assembly Programs . . . 69
  - 4.5.3 Experimental Results . . . 73
- 4.6 Summary . . . 76
- References . . . 77

5 Optimization Techniques for Low Power Circuits — 81
- 5.1 Power Optimization by Transistor Sizing . . . 82
- 5.2 Combinational Logic Level Optimization . . . 84
  - 5.2.1 Path Balancing . . . 84
  - 5.2.2 Don't-care Optimization . . . 85
  - 5.2.3 Logic Factorization . . . 86
  - 5.2.4 Technology Mapping . . . 87
- 5.3 Sequential Optimization . . . 89
  - 5.3.1 State Encoding . . . 89
  - 5.3.2 Encoding in the Datapath . . . 90
  - 5.3.3 Gated Clocks . . . 90
- 5.4 Summary . . . 91

		References	. . . .	92
6	**Retiming for Low Power**			**97**
	6.1	Review of Retiming	. . . .	99
		6.1.1	Basic Concepts	99
		6.1.2	Applications of Retiming	101
	6.2	Retiming for Low Power		101
		6.2.1	Cost Function	102
		6.2.2	Verifying a Given Clock Period	104
		6.2.3	Retiming Constraints	104
		6.2.4	Executing the Retiming	105
	6.3	Experimental Results		107
	6.4	Conclusions		108
		References	. . . .	109
7	**Precomputation**			**111**
	7.1	Subset Input Disabling Precomputation		112
		7.1.1	Subset Input Disabling Precomputation Architecture .	113
		7.1.2	An Example	115
		7.1.3	Synthesis of Precomputation Logic	116
		7.1.4	Multiple-Output Functions	120
		7.1.5	Examples of Precomputation Applied to Datapath Modules	125
		7.1.6	Multiple Cycle Precomputation	126
		7.1.7	Experimental Results for the Subset Input Disabling Architecture	128
	7.2	Complete Input Disabling Precomputation		131
		7.2.1	Complete Input Disabling Precomputation Architecture	132
		7.2.2	An Example	133
		7.2.3	Synthesis of Precomputation Logic	134
		7.2.4	Simplifying the Original Combinational Logic Block .	138
		7.2.5	Multiple-Output Functions	139
		7.2.6	Experimental Results for the Complete Input Disabling Architecture	139
	7.3	Combinational Precomputation		141
		7.3.1	Combinational Logic Precomputation	141
		7.3.2	Precomputation at the Inputs	143
		7.3.3	Precomputation for Arbitrary Sub-Circuits in a Circuit	143

		7.3.4	Experimental Results for the Combinational Precomputation Architecture	146
	7.4	Multiplexor-Based Precomputation		147
	7.5	Conclusions		148
		References		149

8 High-Level Power Estimation and Optimization   151

	8.1	Register Transfer Level Power Estimation		152
		8.1.1	Functional Modules	152
		8.1.2	Controller	157
		8.1.3	Interconnect	158
	8.2	Behavioral Level Synthesis for Low Power		159
		8.2.1	Transformation Techniques	159
		8.2.2	Scheduling Techniques	162
		8.2.3	Allocation Techniques	167
		8.2.4	Optimizations at the Register-Transfer Level	168
	8.3	Conclusions		168
		References		169

9 Conclusion   173

	9.1	Power Estimation at the Logic Level	173
	9.2	Optimization Techniques at the Logic Level	175
	9.3	Estimation and Optimization Techniques at the RT Level . . .	177
		References	178

List of Figures

1.1	A complete synthesis system.	2
2.1	Dynamic vs. static circuits.	14
2.2	Spatial correlation between internal signals.	15
2.3	Computing static probabilities using BDDs.	16
2.4	Glitching due to different input path delays.	17
2.5	Example of a transition waveform.	18
3.1	Example circuit for symbolic simulation.	24
3.2	Symbolic network for a zero delay model.	24
3.3	Possible transitions under a unit delay model.	25
3.4	Symbolic network for a unit delay model.	25
3.5	Pseudo-code for the symbolic simulation algorithm.	27
3.6	Example input waveforms and output waveform for a latch.	28
3.7	Transmission gate and latch.	28
3.8	Example of a combinational circuit with latches.	29
3.9	Symbolic network for a combinational circuit with latches.	29
4.1	A k-pipeline.	36
4.2	Taking k levels of correlation into account.	36
4.3	A synchronous sequential circuit.	37
4.4	Example state transition graph.	38
4.5	Generating temporal correlation of present state lines.	39
4.6	BDD for $ps_1 = ps_2 \vee (I \wedge \overline{ps_1})$.	41
4.7	An m-expanded network with $m = 2$.	52
4.8	Calculation of signal and transition probabilities by network unrolling.	54
4.9	Example circuit: (a) State transition graph; (b) Logic circuit.	55

4.10 (a) Circuit with a redundant state line; (b) 1-unrolled symbolic network. 55
4.11 Symbolic network with $k = 2$. 56
4.12 Example of autonomous IMFSM for a four-vector sequence. . 67
4.13 Cascade of IMFSM and given sequential circuit. 68
4.14 Example of Mealy IMFSM for a four-vector sequence. 69
4.15 Processor model. 71
4.16 Example of Mealy IMFSM for an assembly program. 72
4.17 Generation of transition probabilities: (a) pipeline; (b) cyclic circuit. 78

5.1 Logic restructuring to minimize spurious transitions. 85
5.2 Buffer insertion for path balancing. 85
5.3 SDCs and ODCs in a multilevel circuit. 85
5.4 Logic factorization for low power. 86
5.5 Circuit to be mapped, with switching activity information. . . 87
5.6 Information about the technology library. 87
5.7 Mapping for minimum area. 88
5.8 Mapping for minimum power. 88
5.9 Reducing switching activity in the register file and ALU by gating the clock. 91

6.1 Adding a flip-flop to a circuit. 98
6.2 Moving a flip-flop in a circuit. 98
6.3 Pipelined 2-bit adder: (a) Circuit; (b) Graph. 99
6.4 Retimed 2-bit adder: (a) Circuit; (b) Graph. 100
6.5 Sensitivity calculation. 102
6.6 Vertex selection: (a) Circuit; (b) Binary tree. 105
6.7 Circuit with the gates in the selected set retimed. 106

7.1 Original circuit. 112
7.2 Subset input disabling precomputation architecture. 113
7.3 Subset input disabling precomputation architecture applied to a finite state machine. 114
7.4 A comparator example. 115
7.5 Procedure to determine the optimal subset of inputs to the precomputation logic. 118
7.6 Procedure to determine the optimal set of outputs. 121

7.7	Logic duplication in a multiple-output function.	123
7.8	Procedure to determine a good subset of outputs.	124
7.9	Precomputation applied to a maximum circuit.	125
7.10	Precomputation applied to a carry-select adder.	126
7.11	Multiple cycle precomputation.	127
7.12	Adder-comparator circuit.	128
7.13	Adder-maximum circuit.	129
7.14	Complete input disabling precomputation architecture.	132
7.15	A modified comparator.	133
7.16	Modified comparator under the complete input disabling architecture.	134
7.17	Input selection for the complete input disabling architecture.	136
7.18	Recursive procedure for input selection for the complete input disabling architecture.	137
7.19	Original combinational sub-circuit.	141
7.20	Sub-circuit with input disabling circuit.	142
7.21	Complete input disabling for combinational circuits.	143
7.22	Combinational logic precomputation.	144
7.23	Procedure to find the minimum set of single-output subcircuits.	145
7.24	Precomputation using the Shannon expansion.	148
8.1	Bit transition probability as a function of temporal correlation.	153
8.2	Entropy and switching probability of a Boolean signal as a function of its static probability.	155
8.3	Silage description of the `dealer` circuit.	160
8.4	Control Data Flow Graph for the `dealer` circuit.	161
8.5	Chain vs. tree operations.	161
8.6	Trading a multiplication for an addition.	162
8.7	Control Data Flow Graph for $\|a - b\|$.	163
8.8	Schedule for $\|a - b\|$ using two control steps.	164
8.9	Schedule for $\|a - b\|$ using three control steps.	164
8.10	A power managed schedule for $\|a - b\|$ using three control steps.	165
8.11	CDFG of `dealer` with control edges for power management.	166
8.12	Controller (a)without and (b)with power management.	167

List of Tables

3.1	Statistics of examples.	31
3.2	Power estimation for combinational logic.	31
4.1	Comparison of sequential power estimation methods for pipelined circuits.	57
4.2	Comparison of power estimation methods for cyclic circuits.	59
4.3	Comparison of power estimation methods for cyclic circuits (contd).	60
4.4	Absolute errors in present state line probabilities averaged over all present state lines.	61
4.5	Absolute errors in switching activity averaged over all circuit lines.	62
4.6	Comparison of Picard-Peano and Newton-Raphson.	63
4.7	Results of power estimation using k-unrolled and m-expanded networks.	64
4.8	Percentage error in switching activity estimates averaged over all nodes in the circuit.	65
4.9	α_0 instruction set.	70
4.10	Comparison of power dissipation under uniform input assumption and IMFSM computation.	74
4.11	Comparison of power dissipation under uniform input assumption and IMFSM computation (contd).	75
4.12	Present state line probability errors.	76
6.1	Results of retiming for low power with no timing constraints.	107
6.2	Results of retiming for low power and minimum delay.	108
7.1	Power reductions for datapath circuits.	130

7.2 Power reductions for random logic circuits. 131
7.3 Power reductions in sequential precomputation using the complete input disabling architecture. 139
7.4 Power reductions in sequential precomputation using the complete input disabling architecture (contd). 140
7.5 Comparison of power reductions between complete and subset input disabling architectures. 141
7.6 Power reductions using combinational precomputation. 147

Preface

Traditionally, Computer-Aided Design (CAD) tools have focused on improving performance and reducing area of integrated circuits (ICs), while increasing designer productivity. Beginning about five years ago, the reduction of power dissipation in ICs became increasingly desirable for two reasons, namely reducing heat dissipation in ICs and increasing battery life for portable electronic systems. In response to this desire, CAD research in academia and industry targeted analysis, optimization, and synthesis strategies for low power electronic circuits. Initially, the focus was on layout, transistor-level and logic-level techniques, but quickly the focus expanded to include register-transfer-level, architecture-level, and system-level techniques as well. Over the past 2-3 years, the CAD industry has been marketing products which aid in the reduction of IC power dissipation at various levels of abstraction. These products are integrated into synthesis frameworks that begin from behavioral or structural hardware descriptions of ICs.

The primary focus in this book is on the logic level, which includes combinational and sequential logic. Representative techniques for power analysis and optimization-based synthesis for lower power at the logic level are described in detail. In addition, early CAD research focusing on register-transfer and behavioral levels of abstraction is described in the last chapter.

Acknowledgments

Over the years, several people have helped to deepen our understanding of VLSI synthesis and low power electronic system design. We thank Jonathan Allen, Pranav Ashar, Robert Brayton, Raul Camposano, Anantha Chandrakasan, Tim Cheng, Abhijit Ghosh, Gary Hachtel, Kurt Keutzer, Bill Lin, Sharad Malik, Richard Rudell, Fabio Somenzi and Albert Wang.

We would like to thank Mazhar Alidina for his initial work on pre-computation. Special thanks go to Luís Miguel Silveira for all the support and help. We are appreciative of all the exceptional people on the 8th floor of building 36 who have made it an excellent place to work. Namely thanks to Mike Chou, Mattan Kamon, Stan Liao, Ignacio McQuirk, Ricardo Telichevesky and Jacob White for their friendship and help.

Other people also contributed to make MIT such a great experience. Thanks to Júlia Allen, Miguel and Inês Castro, Jorge Gonçalves, Cristina Lopes, Gail Payne, Nuno and Manuela Vasconcelos.

Lastly, we thank our families for their continual patience and encouragement, especially Tila Moreno, Sheela and Sulochana Devadas, to whom we dedicate this book.

Chapter 1

Introduction

Digital integrated circuits are ubiquitous in systems that require computation. During the years of their inception, the use of integrated circuits was confined to traditional electronic systems such as computers, high-fidelity sound systems, and communication systems. Today not only do computer and communication systems play an increasingly important role, but also the use of integrated systems is much more widespread, from controllers used in washing machines to the automobile industry. As a result, digital circuits are becoming more application specific.

The shrinking of device sizes due to the improvement of fabrication technology has increased dramatically the number of transistors available for use in a single chip. Functions that were performed by several chips can now be done within a single chip, reducing the physical size of the electronic component of the system. The larger capacity of the chips is also being used to extend the functionality of the systems. The overall consequence is a substantial increase in complexity of the integrated circuits.

In order to handle the ever increasing complexity, computer-aided design tools have been developed. These tools have to be general enough to produce good solutions for the broad range of applications for which integrated circuits are being designed.

The first generation of computer-aided design tools dealt with automatically generating the layout masks from the description of the circuit at the logic level. Then logic synthesis tools were introduced to obtain optimized logic circuits from some input/output specification. More recently, tools that can do system-level optimization given a Register-Transfer Level (RTL)

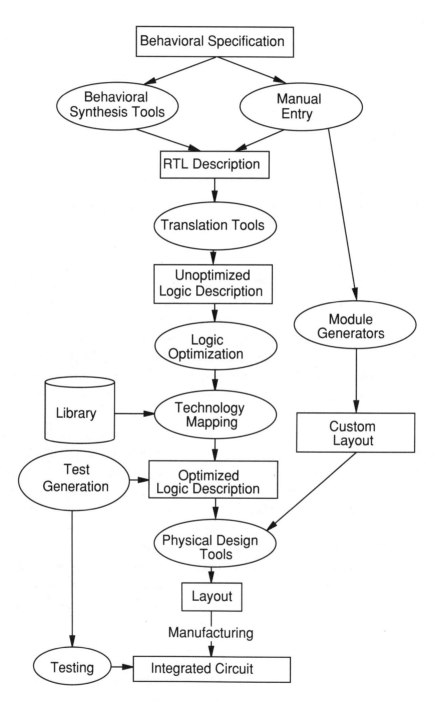

Figure 1.1: A complete synthesis system.

description have been proposed. The trend towards moving the circuit specification to higher description levels continues with research being conducted at the behavioral synthesis level. At this level, the circuit description is akin to an algorithmic description and the synthesis tool decides which registers and functional units to use and assigns each operation to one of these units on a given clock cycle.

A complete synthesis system is presented in Figure 1.1. Each synthesis tool translates a description of the circuit into an optimized description at a lower level. At every description level, area, timing and power dissipation estimates can be obtained and used to drive the synthesis tool such that the design's constraints are met. If at some level any of these constraints is violated, the designer needs to go back one or more description levels and redo the synthesis with different parameters, perhaps relaxing some constraint.

As shown in Figure 1.1, the logic synthesis process is usually split into two different phases. First logic optimization is performed on a Boolean description of the circuit. Technology mapping is then performed on this optimized circuit – this consists of translating the generic Boolean description to logic gates existing in the chosen library. This library is specific to the fabrication process that is going to be used and has precise layout, area and timing information for each gate. Design estimates at this level are therefore more precise than at higher levels.

Also shown in Figure 1.1 is the automatic test generator (ATG) module [1]. At the logic level, this tool generates a set of input patterns that attempts to identify possible circuit malfunctions after fabrication. These input patterns can then be used to test the functionality of the fabricated chips and should be small in number to minimize the test time. Although automatic test generation is seemingly independent from the synthesis process, testability-aware synthesis algorithms can dramatically improve the performance of the test generator.

1.1 Power as a Design Constraint

Traditionally the constraints in the design of an integrated circuit have been timing and area [2, 3, 4, 5]. When designing a circuit, there is usually a performance goal which translates to a maximum duration that any logic signal can take to stabilize after the inputs have stabilized. The second concern is that the circuit should take up as little area as possible since die area has a direct correspondence to cost. Further, this is not a linear relationship as the larger the circuit the more probable it is that there is a fabrication process error in a circuit, lowering circuit yield [8, Chapter 2].

However, the importance of low-power dissipative digital circuits is rapidly increasing. For many consumer electronic applications low average power dissipation is desirable and for certain special applications low power dissipation is of critical importance. For personal communication applications like hand-held mobile telephones, low power dissipation may be the tightest constraint in the design. The battery lifetime may be the decisive factor in the success of the product.

More generally, with the increasing scale of integration and faster clock frequencies, we believe that power dissipation will assume greater importance, especially in multi-chip modules where heat dissipation is one of the biggest problems. Even today, power dissipation is already a significant problem for some circuits. General purpose processors such as the Intel Pentium^{TM} and DEC Alpha^{TM} consume 16W and 30W, respectively. Higher temperatures can affect the circuit's reliability and reduce the lifetime of the system [7]. In order to dissipate the heat that is generated, special packaging and cooling systems have to be used, leading to higher costs.

Optimization for low power can be applied at many different levels of the design hierarchy. The average power dissipation of a circuit, like its area or speed, may be significantly improved by changing the architecture of the circuit [6]. Algorithmic and architectural transformations can trade-off throughput, circuit area, and power dissipation. Furthermore, scaling technology parameters such as supply and threshold voltages can substantially reduce power dissipation. But once these architectural or technological improvements have been made, it is the switching of the logic that will ultimately determine its power dissipation.

In this book we describe a methodology for the optimization of digital circuits for low power at the logic level. The techniques developed are independent of the power reduction techniques applied at higher levels and can be used after system-level decisions are made and high-level transformations applied.

To effectively optimize designs for low power, however, accurate power estimation methods must be developed and used. Power dissipation is generally considered to be more difficult to compute than the estimation of other circuit parameters, like area and delay. The main reason for this difficulty

1. INTRODUCTION 5

is that power dissipation is dependent on the activity of the circuit. In the first part of this book we focus on the power estimation problem.

1.2 Organization of this Book

This book is organized in two main parts. The first part addresses the problem of estimating the average power dissipation of a circuit given its description at the logic level. We start by describing in Chapter 2 the issues involved in computing the power dissipation of a digital circuit. We show that power is directly related to the switching activity of the signals in the circuit. We provide a critique of existing power estimation techniques, namely by pointing out how each technique addresses the issues previously mentioned.

Chapter 3 presents our approach to power estimation for combinational logic circuits. We discuss the merits and drawbacks of our approach and provide comparisons with previous methods.

The power estimation techniques mentioned in Chapters 2 and 3 target combinational circuits. In general, digital integrated circuits are sequential, i.e., they contain memory elements. Chapter 4 describes the technique we have developed that extends the method of Chapter 3 to the sequential circuit case. However, this technique is general enough to be used with any other combinational power estimation method.

One other factor that needs to be taken into account in accurate power estimation is the temporal correlation of primary inputs. Also in Chapter 4, we show how to model this correlation and obtain an accurate power estimation by making use of a sequential power estimator.

The second part of the book is devoted to optimization methods for low power. Chapter 5 presents a survey of the most significant techniques that have been proposed thus far to reduce the power consumption of digital circuits at the logic level.

The next two chapters present original work on sequential logic optimization for low power. Chapter 6 describes a retiming technique for low power. The main observation is that the switching activity at the output of a register can be significantly less than that at the register's input. Any glitching in the input signal is filtered by the register. The technique we propose repositions the registers in the logic circuit such that the overall switching activity in the circuit is minimized.

A power management optimization technique at the logic level is presented in Chapter 7. The logic values at the output of a circuit are selectively precomputed one clock cycle before they are required, and these precomputed values are used to reduce internal switching activity in the succeeding clock cycle. For a large number of circuits, significant power reductions can be achieved by this data-dependent circuit power down technique.

Given the potentially higher impact on power dissipation, and the acquired knowledge at gate level, the focus of research for low power is changing to techniques that work at higher abstraction levels. In Chapter 8 we survey the most significant techniques proposed thus far for power estimation and optimization at the behavioral and RT levels.

Finally, Chapter 9 concludes this book by making a retrospective examination of what has been achieved in the field of power estimation and optimization for low power, and provides directions for future research.

References

[1] M. Abramovici, M. Breuer, and A. Friedman. *Digital Systems Testing and Testable Design*. Computer Science Press, 1990.

[2] P. Ashar, S. Devadas, and A. R. Newton. *Sequential Logic Synthesis*. Kluwer Academic Publishers, Boston, Massachusetts, 1991.

[3] D. Bostick, G. Hachtel, R. Jacoby, M. Lightner, P. Moceyunas, C. Morrison, and D. Ravenscroft. The Boulder Optimal Logic Design System. In *Proceedings of the International Conference on Computer-Aided Design*, pages 62–65, November 1987.

[4] R. Brayton, G. Hachtel, C. McMullen, and A. Sangiovanni-Vincentelli. *Logic Minimization Algorithms for VLSI Synthesis*. Kluwer Academic Publishers, 1984.

[5] R. Brayton, R. Rudell, A. Sangiovanni-Vincentelli, and A. Wang. MIS: A Multiple-Level Logic Optimization System. *IEEE Transactions on Computer-Aided Design*, 6(6):1062–1081, November 1987.

[6] A. Chandrakasan, M. Potkonjak, R. Mehra, J. Rabaey, and R. Broderson. Optimizing Power Using Transformations. *IEEE Transactions on Computer-Aided Design*, 14(1):12–31, January 1995.

[7] A. Christou, editor. *Electromigration and Electronic Device Degradation.* Wiley, 1994.

[8] D. Walker. *Yield Simulation for Integrated Circuits.* Kluwer Academic Publishers, 1987.

Chapter 2

Power Estimation

For power to be used as a design parameter, tools are needed that can efficiently estimate the power consumption of a given design. As in most engineering problems we have tradeoffs, in this case between the accuracy and run-time of the tool.

Accurate power values can be obtained from circuit-level simulators such as SPICE [18]. In practice, these simulators cannot be used in circuits with more than a few thousand transistors, so their applicability in logic design is very limited – they are essentially used to characterize simple logic cells.

A good compromise between accuracy and complexity is switch-level simulation. Simulation of entire chips can be done within reasonable amounts of CPU time [21, 19]. This property makes switch-level simulatorsvery important power diagnosis tools. After layout and before fabrication these tools can be used to identify *hot spots* in the design, i.e., areas in the circuit where current densities or temperature may exceed the safety limits during normal operation.

At the logic level, a more simplified power dissipation model is used, leading to a faster power estimation process. Although detailed circuit behavior is not modeled, the estimation values can still be reasonably accurate. Obtaining fast power estimates is critical in order to allow a designer to compare different designs. Further, for the purpose of directing a designer or a synthesis tool for low power design, rather than an absolute measure of how much power a particular circuit consumes, an accurate relative power measure between two designs will suffice.

This observation is carried out further to justify power estimation schemes at higher abstraction levels. We discuss these methods in Chapter 8.

In this chapter we focus on power estimation at the logic level. This level is perhaps where the best accuracy versus run-time tradeoff is reached. We first describe the power dissipation model that we use at the logic level in Section 2.1. We then present in Section 2.2 a survey of the most significant power estimation techniques at the logic level that have been previously proposed. Both simulation-based (Section 2.2.1) and probabilistic (Section 2.2.3) techniques are reviewed and the issues involved in each technique are discussed.

2.1 Power Dissipation Model

The sources of power dissipation in CMOS devices are summarized by the following expression [24, p. 236]:

$$P = \frac{1}{2} \cdot C \cdot V_{DD}^2 \cdot f \cdot N + Q_{SC} \cdot V_{DD} \cdot f \cdot N + I_{leak} \cdot V_{DD} \qquad (2.1)$$

where P denotes the total power, V_{DD} is the supply voltage, and f is the frequency of operation.

The first term in Equation 2.1 corresponds to the power involved in charging and discharging circuit nodes. C represents the node capacitances and N is the switching activity, i.e., the number of gate output transitions per clock cycle (also known as *transition density* [14]). $\frac{1}{2} \cdot C \cdot V_{DD}^2$ is the energy involved in charging or discharging a circuit node with capacitance C and $f \cdot N$ is the average number of times per second that the nodes switches.

The second term in Equation 2.1 represents the power dissipation due to current flowing directly from the supply to ground during the (hopefully small) period that the pull-up and pull-down networks of the CMOS gate are both conducting when the output switches. This current is often called *short-circuit current*. The factor Q_{SC} represents the quantity of charge carried by the short-circuit current per transition.

The third term in Equation 2.1 is related to the static power dissipation due to leakage current I_{leak}. The transistor source and drain diffusions in a MOS device form parasitic diodes with bulk regions. Reverse bias currents in these diodes dissipate power. Subthreshold transistor currents also dissipate power. I_{leak} accounts for both these small currents.

These three factors for power dissipation are often referred to as *dynamic* power, *short-circuit* power and *leakage current* power respectively.

It has been shown [3] that during normal operation of well designed CMOS circuits the switching activity power accounts for over 90% of the total power dissipation. Thus power optimization techniques at different levels of abstraction target minimal switching activity power. The model for power dissipation for a gate i in a logic circuit is simplified to:

$$P_i = \frac{1}{2} \cdot C_i \cdot V_{DD}^2 \cdot f \cdot N_i \qquad (2.2)$$

The supply voltage V_{DD} and the clock frequency f are defined prior to logic design. The capacitive load C_i that the gate is driving can be extracted from the circuit. This capacitance includes the source-drain capacitance of the gate itself, the input capacitances of the fanout gates and, if available, the wiring capacitance. Therefore the problem of logic level power estimation reduces to computing an accurate estimate of the average number of transitions N_i for each gate in the circuit. In the remainder of this chapter we present a review and critique of techniques for the computation of switching activity in logic circuits.

2.2 Switching Activity Estimation

The techniques we present in this section target *average* switching activity estimation. This is typically the value used to guide optimization methods for low power.

Some work has been done on identifying and computing conditions which lead to *maximum* power dissipation. In [6] a technique is presented that implicitly determines the two input vector sequence that leads to maximum power dissipation in a combinational circuit. More recently, in [13] a method for computing the multiple vector cycle in a sequential circuit that dissipates maximum average power is described.

2.2.1 Simulation-Based Techniques

A straightforward approach to obtain an average transition count at every gate in the circuit is to use a logic or timing simulator and simulate the circuit for a *sufficiently large* number of randomly generated input vectors. The main advantage of this approach is that existing logic simulators can be used directly and issues such as glitching and internal signal correlation are automatically taken into account by the logic simulator.

The most important aspect of simulation-based switching activity estimation is deciding how many input vectors to simulate in order to achieve a given accuracy level. A basic assumption is that under random inputs the power consumed by a circuit over a period of time T has a Normal distribution. Given a user-specified allowed percentage error ϵ and confidence level α, the approach described in [2] uses the Central Limit Theorem [16, pp. 214-221] to compute the number of input vectors with which to simulate the circuit with. With $\alpha \times 100\%$ confidence, $|\bar{p} - P| < \text{erf}^{-1}(\frac{\alpha}{2}) \times s/\sqrt{L}$, where $\bar{p}$ and s are the measured average and standard deviation of the power, P is the true average power dissipation, L the number of input vectors and $\text{erf}^{-1}(\frac{\alpha}{2})$ is the *inverse error function* [16, p. 49] obtained from the Normal distribution. Since we require $\frac{|\bar{p}-P|}{\bar{p}} < \epsilon$, it follows that

$$L \geq \left(\frac{\text{erf}^{-1}(\frac{\alpha}{2}) \times s}{\epsilon \times \bar{p}} \right)^2 \tag{2.3}$$

For a typical logic circuit and reasonable error and confidence level, the numbers of vectors needed is usually small, making this approach very efficient.

A limitation of the technique presented in [2] is that it only guarantees accuracy for the *average* switching activity over all the gates. The switching activity values for individual gates (N_i in Equation 2.2) may have large errors and these values are important for many optimization techniques.

This method is augmented in [25] by allowing the user to specify the percentage error and confidence level for the switching activity of individual gates. Equation 2.3 is used for each node in the circuit, where instead of power, the average and standard deviation of the number of transitions in the node is the relevant parameter. The number of input vectors L is obtained as the minimum L that verifies Equation 2.3 for all the nodes.

The problem now is that gates which have a low switching probability, *low-density nodes*, may require a very large number of input vectors in order for the estimation to be within the percentage error specified by the user. The authors solve this problem by being less restrictive for these gates: an absolute error bound is used instead of the percentage error. The impact of possible larger errors for low-density nodes is minimized by the fact that these gates have the least effect on power dissipation and circuit reliability.

Other methods [10] try to compute a tighter bound on the number of input vectors to simulate. Instead of relying on normal distribution properties, the authors assume that the number of transitions at the output of a gate has a multinomial distribution. However, this method has to make a number of empirical approximations in order to obtain the number of input vectors.

Simulation-based techniques can be very efficient for loose accuracy bounds. Increasing the accuracy may require a prohibitively high number of simulation vectors. Using simulation-based methods in a synthesis scenario, where a circuit is being incrementally modified and power estimates have to be obtained repeatedly for subsets of nodes in the circuit, can be quite inefficient.

2.2.2 Issues in Probabilistic Estimation Techniques

Given some statistical information of the inputs, probabilistic methods propagate this information through the logic circuit obtaining statistics about the switching activity at each node in the circuit. Only one pass through the circuit is needed making these methods potentially very efficient. Still, modeling issues like correlation between signals can make these methods computationally expensive.

Temporal Correlation: Static vs. Transition Probabilities

The *static* probability of a logic signal x is the probability of x being 0 or 1 at any instant (we will represent this, respectively, as $prob(\bar{x})$ and $prob(x)$). *Transition* probabilities are the probability of x making a 0 to 1 or 1 to 0 transition, staying at 0 or staying at 1 between two time instants. We will represent these probabilities as $prob^{01}(x)$, $prob^{10}(x)$, $prob^{00}(x)$ and $prob^{11}(x)$, respectively. Note that we always have $prob^{01}(x) = prob^{10}(x)$.

The probability of signal x making a transition is given by $prob^{01}(x) + prob^{10}(x)$. Relating to Equation 2.2, $N_x = prob^{01}(x) + prob^{10}(x)$.

Static probabilities can always be derived from transition probabilities:

$$\begin{aligned} prob(x) &= prob^{11}(x) + prob^{01}(x) \\ prob(\bar{x}) &= prob^{00}(x) + prob^{10}(x) \end{aligned} \quad (2.4)$$

Derivation in the other direction is only possible if we are given the correlation coefficients between successive values of a signal. If we assume

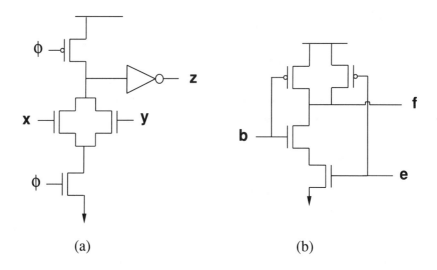

Figure 2.1: Dynamic vs. static circuits.

these values are independent then:

$$\begin{aligned} prob^{11}(x) &= prob(x) \times prob(x) \\ prob^{10}(x) &= prob(x) \times prob(\bar{x}) \\ prob^{01}(x) &= prob(\bar{x}) \times prob(x) \\ prob^{00}(x) &= prob(\bar{x}) \times prob(\bar{x}) \end{aligned} \quad (2.5)$$

In the case of dynamic precharged circuits, exemplified in Figure 2.1(a), the switching activity is uniquely determined by the applied input vector. If both x and y are 0, then z stays at 0 and there is no switching activity. If one or both of x and y are 1, then z goes to 1 during the evaluation phase and back to 0 during precharging. Therefore, the switching activity at z will be twice the *static* probability of z being 1, $N_z = 2 \times prob(z)$.

On the other hand, the switching activity in static CMOS circuits is a function of a two input vector sequence. For instance, consider the circuit shown in Figure 2.1(b). In order to determine if the output f switches we need to know what value it assumed for the first input vector and to what value it evaluated after the second input vector. Using static probabilities one can compute the probability that f evaluates to 1 for the first ($prob_1(f)$) and second ($prob_2(f)$) input vectors. Then:

$$\begin{aligned} N_f &= prob_1(f) \times prob_2(\bar{f}) + prob_1(\bar{f}) \times prob_2(f) \\ &= prob(f) \times (1 - prob(f)) + (1 - prob(f)) \times prob(f) \end{aligned}$$

2. POWER ESTIMATION

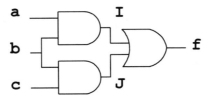

Figure 2.2: Spatial correlation between internal signals.

$$= 2 \times prob(f) \times (1 - prob(f))$$

since $prob_1(f) = prob_2(f) = prob(f)$ and $prob(\bar{f}) = 1 - prob(f)$.

By using static probabilities in the previous expression we ignored any correlation between the two vectors in the input sequence. In general ignoring this type of correlation, called *temporal correlation*, is not a valid assumption. Probabilistic estimation methods work with transition probabilities at the inputs, thus introducing the necessary correlation between input vectors:

$$N_f = prob^{01}(f) + prob^{10}(f).$$

Transition probabilities are computed and propagated for all the nodes in the circuit.

Spatial Correlation

Another type of signal correlation in logic circuits is *spatial correlation*. The probability of two or more signals being 1 may not be independent. Spatial correlation of input signals, even if known, can be difficult to specify, so most probabilistic techniques assume the inputs to be spatially independent. In Section 4.5 we propose a method that takes into account input signal correlation for user-specified input sequences.

Even if spatial independence is assumed for input signals, logic circuits with reconvergent fanout introduce spatial correlation between internal signals. Consider the circuit depicted in Figure 2.2. Assuming that inputs a, b and c are uncorrelated, the static probability at I is $prob(I) = prob(a)prob(b)$ and at J is $prob(J) = prob(b)prob(c)$. However, $prob(f) \neq prob(I) + prob(J) - prob(I)prob(J)$ because I and J are correlated ($b=0 \Rightarrow I=J=0$).

To compute accurate signal probabilities, we need to take into account this internal spatial correlation. One solution to this problem is to write

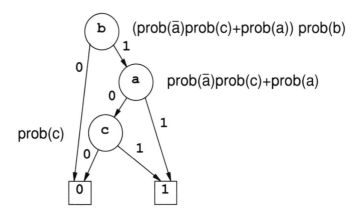

Figure 2.3: Computing static probabilities using BDDs.

the Boolean function as a disjoint sum-of-products expression, where each product-term has a null intersection with any other. For the previous example, we write f as:

$$\begin{aligned} f &= (a \wedge b) \vee (b \wedge c) \\ &= (a \wedge b) \vee (\bar{a} \wedge b \wedge c) \end{aligned}$$

Then $prob(f) = prob(a)prob(b) + prob(\bar{a})prob(b)prob(c)$.

A more efficient approach is to use Binary Decision Diagrams (BDDs) [1]. The static probabilities can be computed in time linear in the size of the BDD by traversing the BDD from leaves to root, since the BDD implements a disjoint cover with sharing. The BDD for the previous example is illustrated in Figure 2.3.

Glitching

Yet another issue is spurious transitions (or glitching) at the output of a gate due to different input path delays. These may cause the gate to switch more than once during a clock cycle, as exemplified in Figure 2.4. Studies have shown that glitching cannot be ignored as it can be a significant fraction of the total switching activity [20, 8].

2.2.3 Probabilistic Techniques

There has been a great deal of work in the area of probabilistic power estimation in the past few years. We describe representative techniques in this

2. POWER ESTIMATION

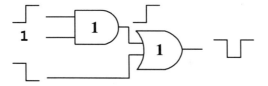

Figure 2.4: Glitching due to different input path delays.

section. These techniques focus on static CMOS circuits since computing transition probabilities is more complex than computing static probabilities. Static probabilities can be obtained from the transition probabilities by Equation 2.4.

Early methods to approximate signal probability targeted testability applications [17, 9, 12, 7]. These methods are not directly applicable to the power estimation problem.

The first approach that was concerned with switching activity for power dissipation was presented in [5]. The static probabilities of the input signals are propagated through the logic gates in the circuit. In this straightforward approach, a zero delay model is assumed, thus glitching is not computed. Since static probabilities are used no temporal signal correlation is taken into account. Further, spatial correlation is also ignored as signals at the input of each gate are assumed to be independent.

In [14], a technique is presented that propagates *transition densities* ($D(x)$) through the circuit. The author shows that the transition density at the output f of a logic gate with n *uncorrelated* inputs x_i can be computed as

$$D(f) = \sum_{i=1}^{n} prob\left(\frac{\partial f}{\partial x_i}\right) D(x_i). \qquad (2.6)$$

$\frac{\partial f}{\partial x_i}$ are the combinations for which the value of f depends on the value of x_i and is given by

$$\frac{\partial f}{\partial x_i} = f_{x_i} \oplus f_{\overline{x_i}}, \qquad (2.7)$$

where $\oplus$ stands for the exclusive-or operator and f_{x_i} and $f_{\overline{x_i}}$ are the cofactors of f with respect to x_i and $\overline{x_i}$, respectively (the cofactors can be obtained simply by setting x_i to a 1 or 0 in f). That is, the switching activity at the output is the sum of the switching activity of each input weighted by the probability that a transition at this input is propagated to the output.

Implicit to this technique is also a zero delay model. An attempt to take glitching into account is suggested by decoupling delays from the logic

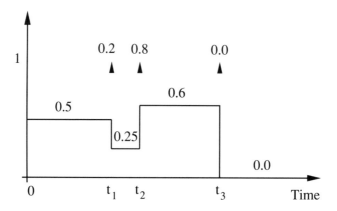

Figure 2.5: Example of a transition waveform.

gate and computing transition densities at each different time point where inputs may switch.

A major shortcoming of this method is the assumption of spatial independence of the input signals to each gate. [11] extends the work of [14] by partially solving this spatial correlation problem. The logic circuit is partitioned in order to compute accurate transition densities at some nodes in the circuit. For each partition, spatial correlation is taken into account by using BDDs.

A similar technique, introduced in [15], uses the notion of *transition waveform*. A transition waveform, illustrated in Figure 2.5, represents an average of all possible signal waveforms at a given node. The example of Figure 2.5 shows that there are no transitions between instants 0 and t_1 and that during this interval half of the possible waveforms are at 1. At instant t_1 a fraction of 0.2 of the waveforms make a 0 to 1 transition, leaving a quarter of the waveforms at 1 (which implies that a fraction of 0.45 of the waveforms make a 1 to 0 transition). A transition waveform basically has all the information about static and transition probabilities of signals and how these probabilities change in time. Their main advantage is to allow an efficient computation of glitching. Transition waveforms are propagated through the logic circuit in much the same way as transition densities.

Again, transition waveform techniques are not able to handle spatial correlations. Another method based on transition waveforms is proposed in [22] where *correlation coefficients* between internal signals are computed beforehand and then used when propagating the transition waveforms. These coefficients are computed for pairs of signals (from their logic AND) and are

2. POWER ESTIMATION 19

based on steady state conditions. This way some spatial correlation is taken into account.

Recent work [4] generalizes the Parker-McCluskey method [17] (a probabilistic technique for testability applications) to handle transition probabilities by using four-valued variables rather than Boolean variables. The Parker-McCluskey method generates a polynomial that represents the probability that the gate output is a 1, as a function of the static probabilities of the primary inputs. It follows basic rules for propagating polynomials through logic gates. The method proposed in [4] can be used to obtain exact (in the sense that temporal and spatial correlation are accurately modeled) switching activities for the zero delay model, but no generalization to handle gate delays was made.

The Boolean Approximation method [23] uses Taylor series expansions to efficiently compute signal probabilities and switching activities. This method is also restricted to the zero delay model. Given two functions A and B, the value computed for $prob(A \wedge B)$ by this method may be in error by as much as 50%, if A and B share more than one input and only the first term in the Taylor series is used. Using higher order Taylor series terms results in much greater complexity.

In Chapter 3, we describe a switching activity estimation technique that follows a different approach and which can effectively handle all the issues mentioned in Section 2.2.2.

2.3 Summary

Power estimation issues and techniques at the logic level have been reviewed. The model used at this abstraction level is such that the power dissipated at the output of a gate is directly proportional to the switching probability of the node. Therefore the problem of power estimation reduces to one of signal probability evaluation.

There are two main approaches for computing the switching activity in a logic circuit: simulation-based and probabilistic techniques. In both the tradeoff is accuracy vs. run-time. In simulation-based methods, the higher the accuracy requested by the user (translated in terms of lower allowed error ϵ and/or higher confidence level α) the more input vectors that have to be simulated. In probabilistic methods, we have methods such as the transition density propagation method [14] that are very fast but ignore some important issues like spatial correlation, to methods such as the extension to Parker-McCluskey [4] that model correlation but are much slower and limited in the size of circuits that can be handled.

References

[1] R. Bryant. Graph-Based Algorithms for Boolean Function Manipulation. *IEEE Transactions on Computers*, C-35(8):677–691, August 1986.

[2] R. Burch, F. Najm, P. Yang, and T. Trick. A Monte Carlo Approach to Power Estimation. *IEEE Transactions on VLSI Systems*, 1(1):63–71, March 1993.

[3] A. Chandrakasan, T. Sheng, and R. Brodersen. Low Power CMOS Digital Design. *Journal of Solid State Circuits*, 27(4):473–484, April 1992.

[4] D. Cheng. *Power Estimation of Digital CMOS Circuits and the Application to Logic Synthesis for Low Power*. PhD thesis, University of California at Santa Barbara, December 1995.

[5] M. Cirit. Estimating Dynamic Power Consumption of CMOS Circuits. In *Proceedings of the International Conference on Computer-Aided Design*, pages 534–537, November 1987.

[6] S. Devadas, K. Keutzer, and J. White. Estimation of Power Dissipation in CMOS Combinational Circuits Using Boolean Function Manipulation. *IEEE Transactions on Computer-Aided Design*, 11(3):373–383, March 1992.

[7] S. Ercolani, M. Favalli, M. Damiani, P. Olivo, and B. Ricco. Testability Measures in Pseudo-Random Testing. *IEEE Transactions on Computer-Aided Design*, 11(6):794–799, June 1992.

[8] M. Favalli and L. Benini. Analysis of Glitch Power Dissipation in CMOS ICs. In *Proceedings of the International Symposium on Low Power Design*, pages 123–128, April 1995.

[9] H. Goldstein. Controllability/Observability of Digital Circuits. *IEEE Transactions on Circuits and Systems*, 26(9):685–693, September 1979.

[10] A. Hill and S. Kang. Determining Accuracy Bounds for Simulation-Based Switching Activity Estimation. In *Proceedings of the International Symposium on Low Power Design*, pages 215–220, April 1995.

[11] B. Kapoor. Improving the Accuracy of Circuit Activity Measurement. In *Proceedings of the International Workshop on Low Power Design*, pages 111–116, April 1994.

[12] B. Krishnamurthy and I. G. Tollis. Improved Techniques for Estimating Signal Probabilities. *IEEE Transactions on Computers*, 38(7):1041–1045, July 1989.

[13] S. Manne, A. Pardo, R. Bahar, G. Hachtel, F. Somenzi, E. Macii, and M. Poncino. Computing the Maximum Power Cycles of a Sequential Circuit. In *Proceedings of the 32nd Design Automation Conference*, pages 23–28, June 1995.

[14] F. Najm. Transition Density: A New Measure of Activity in Digital Circuits. *IEEE Transactions on Computer-Aided Design*, 12(2):310–323, February 1993.

[15] F. Najm, R. Burch, P. Yang, and I. Hajj. Probabilistic Simulation for Reliability Analysis of CMOS VLSI Circuits. *IEEE Transactions on Computer-Aided Design*, 9(4):439–450, April 1990.

[16] A. Papoulis. *Probability, Random Variables and Stochastic Processes*. McGraw-Hill, 3rd edition, 1991.

[17] K. Parker and E. McCluskey. Probabilistic Treatment of General Combinational Networks. *IEEE Transactions on Electronic Computers*, C-24(6):668–670, 1975.

[18] T. Quarles. The SPICE3 Implementation Guide. Technical Report ERL M89/44, Electronics Research Laboratory Report, University of California at Berkeley, Berkeley, California, April 1989.

[19] A. Salz and M. Horowitz. IRSIM: An Incremental MOS Switch-Level Simulator. In *Proceedings of the 26th Design Automation Conference*, pages 173–178, June 1989.

REFERENCES

[20] A. Shen, S. Devadas, A. Ghosh, and K. Keutzer. On Average Power Dissipation and Random Pattern Testability of Combinational Logic Circuits. In *Proceedings of the International Conference on Computer-Aided Design*, pages 402–407, November 1992.

[21] R. Tjarnstrom. Power Dissipation Estimate by Switch Level Simulation. In *Proceedings of the International Symposium on Circuits and Systems*, pages 881–884, May 1989.

[22] C-Y. Tsui, M. Pedram, and A. Despain. Efficient Estimation of Dynamic Power Dissipation under a Real Delay Model. In *Proceedings of the International Conference on Computer-Aided Design*, pages 224–228, November 1993.

[23] T. Uchino, F. Minami, T. Mitsuhashi, and N. Goto. Switching Activity Analysis using Boolean Approximation Method. In *Proceedings of the International Conference on Computer-Aided Design*, pages 20–25, November 1995.

[24] N. Weste and K. Eshraghian. *Principles of CMOS VLSI Design*. Addison-Wesley Publishing Company, second edition, 1994.

[25] M. Xakellis and F. Najm. Statistical Estimation of the Switching Activity in Digital Circuits. In *Proceedings of the 31st Design Automation Conference*, pages 728–733, June 1994.

Chapter 3

A Power Estimation Method for Combinational Circuits

In this chapter we describe a technique for the power estimation of logic circuits. This technique is based on *symbolic simulation* and was first presented in [3]. A variable delay model is used for combinational logic in the symbolic simulation method, which correctly computes the Boolean conditions that cause *glitching* (multiple transitions at a gate) in the circuit. In some cases, glitching may account for a significant percentage of the switching activity [5, 2]. For each gate in the circuit, symbolic simulation produces a set of Boolean functions that represent the conditions for switching at different time points. Given input switching rates, we can use exact or approximate methods to compute the probability of each gate switching at any particular time point. We then sum these probabilities over all the gates to obtain the expected switching activity in the entire circuit over all the time points corresponding to a clock cycle. This method takes into account correlation caused at internal gates in the circuit due to reconvergence of input signals (reconvergent fanout).

We describe the symbolic simulation algorithm in Section 3.1. In Sections 3.2 and 3.3 we show how the symbolic simulation can be used to handle transmission gates and inertial delays, respectively. We present power estimation results for some circuits in Section 3.4.

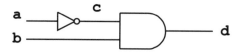

Figure 3.1: Example circuit for symbolic simulation.

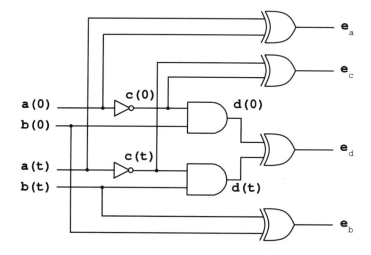

Figure 3.2: Symbolic network for a zero delay model.

3.1 Symbolic Simulation

We build a symbolic network from the symbolic simulation of the original logic circuit over a two input vector sequence. The symbolic network is a logic circuit which has the Boolean conditions for all values that each gate in the original network may assume at different time instants given this input vector pair.

If a zero delay model is used, each gate in the circuit can only assume two different values, one corresponding to each input vector. For this simple case, the symbolic network corresponds to two copies of the original network, one copy evaluated with the first input vector and the other copy with the second. Then, exclusive-or (XOR) gates are added between each pair of nodes that correspond to the same node in the original circuit. The output of an XOR evaluating to a 1 indicates that for this input vector pair the corresponding node in the original circuit makes one transition (it evaluates to a different value for each of the two input vectors).

3. A POWER ESTIMATION METHOD FOR COMBINATIONAL CIRCUITS   25

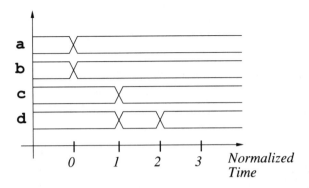

Figure 3.3: Possible transitions under a unit delay model.

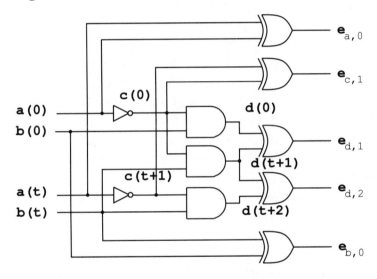

Figure 3.4: Symbolic network for a unit delay model.

To illustrate this process, consider the circuit of Figure 3.1. The symbolic network for a zero delay model is shown in Figure 3.2. The inputs $a(0)$ and $b(0)$ correspond to the first input vector and $a(t)$ and $b(t)$ to the second. If the output e_c evaluates to 1, then signal c in the original circuit (cf. Figure 3.1) will make a transition for the applied vector pair. Similarly for outputs e_a, e_b and e_d.

In the case of unit or general delay models, the gate output nodes of a multilevel network can have multiple transitions in response to a two-vector input sequence. Figure 3.3 shows the possible transitions that the output of each gate in the circuit of Figure 3.1 can make under a unit delay model.

The symbolic simulator is able to simulate circuits with arbitrary gate transport delays. The symbolic network will have nodes corresponding to all intermediate values that each gate in the original circuit may assume. The XOR gates will be connected to nodes corresponding to consecutive time instants and relating to the same node in the original circuit.

The symbolic network for a unit delay model for the circuit of Figure 3.1 is presented in Figure 3.4. Nodes $c(0)$ and $d(0)$ are the initial values of nodes c and d respectively. At instant 1, node c will have the value $c(t+1)$ and d the value $d(t+1)$. $e_{c,1} = c(0) \oplus c(t+1)$ evaluates to 1 only if node c makes a transition at instant 1. Similarly for node d at instant 1. At instant 2, node d will assume the value $d(t+2)$. Again $e_{d,2} = d(t+1) \oplus d(t+2)$ gives the condition for d to switch at instant 2. The total switching at the output of gate d will be the sum of $e_{d,1}$ and $e_{d,2}$.

The pseudo-code for the symbolic simulation algorithm is presented in Figure 3.5. The simulator processes one gate at a time, moving from the primary inputs to the primary outputs of the circuit. For each gate g_i, an ordered list of the possible transition times of its inputs is first obtained. Then, possible transitions at the output of the gate are derived, taking into account transport delays from each input to the gate output. The processing done is similar to the "time-wheel" in a timing simulator.

Once the symbolic network of a circuit is computed, we use the static probabilities of the inputs to obtain the static probabilities of the output of the XORs evaluating to 1. This probability is the same as the switching probability of the nodes in the original circuit.

This method models glitching accurately and if BDDs are used to compute the static probabilities, exact spatial correlation is implicitly taken into account. Temporal correlation of the inputs can be handled during the BDD traversal by using the probabilities of pairs of corresponding inputs, e.g., $\langle a(0),a(t) \rangle$, which are the transition probabilities.

In some cases, the BDDs for the generated functions may be too large. The signal probability calculation can be done by a process of random logic simulation. A large number of randomly generated vectors are simulated on the symbolic network till the signal probability value converges to within 0.1%. Levelized/event-driven simulation methods that simulate 32 vectors at a time can be used in an efficient probability evaluation scheme. The probabilities thus obtained are statistical approximations.

```
1.      Gates = Topological_Sort( Network ) ;
2.      for each g_i in Gates {
3.          if g_i is a primary input then {
4.              TimePoints = { (0, f_i(0)), (t, f_i(t)) } ;
5.              e_{i,t} = f_i(0) ⊕ f_i(t) ;
6.          }
7.          else {
8.              Δ = delay of g_i ;
9.              TimePoints = NIL(LIST) ;
10.             for each input g_j of g_i ( g_{i_1}, ···, g_{i_m} ) {
11.                 for each time point (k, f_j(k)) of g_j {
12.                     TimePoints = InsertInOrder ( TimePoints, (k, f_j(k)) ) ;
13.                 }
14.             }
15.             /* g̃_i is the Boolean function of gate g_i with respect to
16.                its immediate inputs */
17.             f_i(0) = g̃_i(f_{i_1}(0), ··· f_{i_m}(0)) ;
18.             l = 0 ;
19.             for each new time point k in TimePoints {
20.                 f_i(k + Δ) = g̃_i(f_{i_1}(k), ··· f_{i_m}(k)) ;
21.                 e_{i,k+Δ} = f_i(l) ⊕ f_i(k + Δ) ;
22.                 l = k + Δ ;
23.             }
24.         }
25.     }
```

Figure 3.5: Pseudo-code for the symbolic simulation algorithm.

3.2 Transparent Latches

We describe how symbolic simulation handles combinational circuits with embedded transparent latches or transmission gates.

Transmission gates have an input, an output, and a control line, as depicted in Figure 3.6. When the control line is high, the output is identical to the input. When the control line is low, however, the output is given by

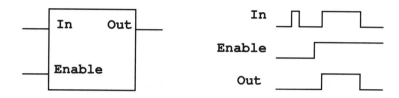

Figure 3.6: Example input waveforms and output waveform for a latch.

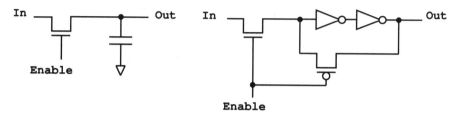

Figure 3.7: Transmission gate and latch.

value stored in the previous time instant. Examples of a transmission gate and a transparent latch are shown in Figure 3.7.

It is this feature of having memory that makes transmission gates different from normal combinational gates like an AND gate. In mathematical terms, if a is the input, b the control, and x the output, then at any time instant t, the output of a transmission gate is given as

$$x(t) = b(t) \wedge a(t) \vee \overline{b(t)} \wedge x(t-1), \qquad (3.1)$$

where $t - 1$ refers to the previous time instant.

From the switching activity estimation viewpoint, the symbolic simulation approach handles transmission gates (or transparent latches) in a straightforward manner. Since $x(t-1)$ is computed before $x(t)$ in the simulation, we create functions corresponding to the different $x(t)$'s and use them in simulating the fanout gates. We use the symbolic input $b(t)$ during symbolic simulation of $x(t)$. As the symbolic simulation proceeds, the known equations for the time points for each input are used and the logic equations corresponding to the various transitions at the output of the latch are computed. As a result, in a single pass from inputs to outputs, switching activity estimation can be carried out for an acyclic circuit.

If the initial value $x(-1)$ (the value of x before the first input vector is applied) is known it is replaced by the appropriate 0 or 1 value during symbolic

3. A POWER ESTIMATION METHOD FOR COMBINATIONAL CIRCUITS

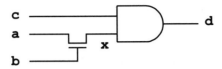

Figure 3.8: Example of a combinational circuit with latches.

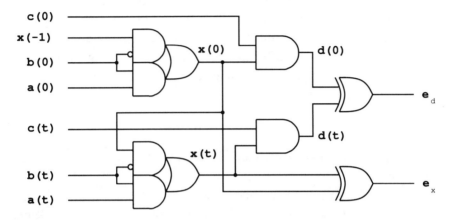

Figure 3.9: Symbolic network for a combinational circuit with latches.

simulation. If the initial value $x(-1)$ is not known, it can be replaced by a Boolean variable with a signal probability of 0.5.

To illustrate the symbolic simulation process of a transmission gate, consider the simple circuit depicted in Figure 3.8. The symbolic network for this circuit assuming a zero delay model is shown in Figure 3.9 (to simplify the picture, the XOR's for the primary inputs are not shown). The difference between this symbolic network and the one for a combinational circuit is that we have a logic signal corresponding to a previous time instant $x(0)$ feeding a gate that generates the same signal for the next time instant $x(t)$.

3.3 Modeling Inertial Delay

Logic gates require energy to switch state. The energy in an input signal to a gate is a function of its amplitude and duration. If its duration is too small, the signal will not force the gate to switch. The minimum duration for which an input change must persist in order for the gate to switch states is called the *inertial delay* of an element and is denoted by Δ (cf. [1, p. 187]).

Inertial delay is usually modeled at the inputs to gates. However, for our purposes it is more convenient to model it at the gate output. We will assign an integer $\Delta_i \geq 0$ to each gate i. Δ_i is obtained from process and device parameters like propagation delay. We then require that any pair of output transitions at i be separated by at least a duration Δ_i.

The symbolic simulation proceeds as described in the previous sections to compute $f_i(t), ..., f_i(t+l)$. If we have $\Delta_i > 0$, then if there is a transition between time t and $t+1$ we cannot have a transition between $t+1$ and $t+\Delta_i$. Therefore, if we have three different time points, $f_i(t_1)$, $f_i(t_2)$ and $f_i(t_3)$, within Δ_i from t_1 we make sure there are no transitions by making $f_i(t_2) = f_i(t_1)$ when $f_i(t_1) = f_i(t_3)$. We create

$$f'_i(t_2) = f_i(t_2) \wedge (f_i(t_1) \vee f_i(t_3)) \vee (f_i(t_1) \wedge f_i(t_3)) \quad (3.2)$$

for every three time points within Δ_i. We compute $f'_i(t_3)$ using $f'_i(t_2)$ and $f_i(t_4)$ and so on.

The $f'_i(t)$ functions are used as the inputs to the XOR gates to compute the switching activities. Also, we use the $f'_i(t)$ functions for the next logic level, thus any transitions eliminated at the output of a gate are not propagated to its transitive fanout.

3.4 Power Estimation Results

Throughout this section, we will be measuring the average power dissipation of the circuit by using Equation 2.2 summed over all the gates in the circuit. The N_i values are computed for the gates in the circuit under different delay models. Since the circuits are technology-mapped circuits, the load capacitance values of the gates are known. A clock frequency of 20MHz and supply voltage of 5V have been assumed. The power estimates are given in micro-Watt.

The statistics of the examples used are shown in Table 3.1. All of the examples except the last two belong to the ISCAS-89 Sequential Benchmark set. Example `add16` is a 16-bit adder and `max16` is a 16-bit maximum function.

All the circuits considered are technology-mapped static CMOS circuits. For all the circuits, we assumed uniform static (0.5) and transition (0.25) probabilities for the primary inputs. Note, however, that user-provided non-uniform probabilities could just as easily been used.

3. A POWER ESTIMATION METHOD FOR COMBINATIONAL CIRCUITS

Circuit	Inputs	Outputs	Regs	Gates
s27	4	1	3	10
s298	3	6	14	119
s349	9	11	15	150
s386	7	7	6	159
s420	19	2	16	196
s510	19	7	6	211
s641	35	24	19	379
s713	35	23	19	393
s838	35	2	32	390
s1238	14	14	18	508
s1494	8	19	6	647
add16	33	17	16	288
max16	33	16	16	154

Table 3.1: Statistics of examples.

Circuit	Zero Delay Power	Unit Delay Power	Variable Delay Power	CPU Time BDD	CPU Time Logic
s27	82	93	93	0.1	0.2
s298	922	1033	1069	5.2	2.3
s349	777	1094	1110	9.7	6.1
s386	1070	1183	1250	9.2	4.9
s420	877	940	958	12.0	5.2
s510	993	1236	1331	11.2	5.5
s641	1228	1594	1665	62.6	36.3
s713	1338	1847	1932	151.6	92.6
s838	1727	1822	1847	52.6	16.9
s1238	2394	3013	3158	115.1	43.7
s1494	3808	4762	5045	68.9	32.2
add16	1258	1725	1741	10.4	6.1
max16	599	713	713	4.2	1.6

Table 3.2: Power estimation for combinational logic.

We focus on estimating switching activity and power dissipation in the *combinational logic* of the given circuits. In Table 3.2, the effects of various delay models on the power estimate are illustrated. In the zero delay model, all gates have zero delay and therefore they switch instantaneously. In the unit delay model, all gates have one unit delay. Using the zero delay model ignores glitches in the circuit, and therefore power dissipation due to glitches is not taken into account. The unit delay model takes into account glitches, but a constant delay value is assumed for all gates. The variable delay model uses different delays for different gates, thus is the most realistic model.

Only the times required to obtain the power estimate for the variable delay model are shown in the last column. The variable delay computations are the most complex and therefore power estimation under this model takes the most time. The CPU times correspond to a DEC 3000/900 with 256Mb of memory, and are in seconds. The signal probability calculation was done using two different methods. The column BDD corresponds to exact signal probability evaluation of the output of the XOR gates of Section 3.1 using ordered Binary Decision Diagrams.

Using random logic simulation to evaluate signal probabilities required substantially less CPU time for the large examples as shown in the column LOGIC. Random logic simulation was carried out until the signal probability of each XOR output converged to within 0.1%. This required the simulation of between 1000-50,000 vectors for the different examples. The power measures obtained using the two methods BDD and LOGIC are identical.

3.5 Summary

We presented an algorithm that probabilistically estimates the switching activity in combinational logic circuits. Results indicate that this algorithm is applicable to circuits of moderate size. The most desirable feature of this algorithm is that correlation between internal signals is implicitly taken into account under a variable delay model. Additionally, glitching at any node in the circuit is accurately modeled. Given the delay model chosen, the BDD-based method of estimating switching activity is exact.

Further, we presented an extension of the symbolic simulation algorithm to model transparent latches and inertial delays.

In order to perform exact signal probability evaluation, ordered Binary Decision Diagrams have to be used. Ordered BDDs cannot be built for large multipliers ($\geq$ 16 bits) and for very large circuits. Approximate techniques have to be used in these cases. Our experience with random logic simulation for signal probability evaluation has been favorable.

The symbolic simulation package has been implemented within SIS [4], the synthesis environment from the CAD group at the University of California at Berkeley, and is now part of their standard distribution.

Correlation between primary inputs exists when a given combinational circuit is embedded in a larger sequential circuit. The techniques described have to be augmented to handle sequential circuits and primary input correlation. These issues are dealt with in the next chapter.

References

[1] M. Breuer and A. Friedman. *Diagnosis and Reliable Design of Digital Systems*. Computer Science Press, 1976.

[2] M. Favalli and L. Benini. Analysis of Glitch Power Dissipation in CMOS ICs. In *Proceedings of the International Symposium on Low Power Design*, pages 123–128, April 1995.

[3] A. Ghosh, S. Devadas, K. Keutzer, and J. White. Estimation of Average Switching Activity in Combinational and Sequential Circuits. In *Proceedings of the 29th Design Automation Conference*, pages 253–259, June 1992.

[4] E. Sentovich, K. Singh, C. Moon, H. Savoj, R. Brayton, and A. Sangiovanni-Vincentelli. Sequential Circuit Design Using Synthesis and Optimization. In *Proceedings of the International Conference on Computer Design: VLSI in Computers and Processors*, pages 328–333, October 1992.

[5] A. Shen, S. Devadas, A. Ghosh, and K. Keutzer. On Average Power Dissipation and Random Pattern Testability of Combinational Logic Circuits. In *Proceedings of the International Conference on Computer-Aided Design*, pages 402–407, November 1992.

Chapter 4

Power Estimation for Sequential Circuits

The power estimation methods described in the previous chapters apply to combinational logic blocks. In this chapter we describe techniques that target issues particular to sequential circuits.

While for combinational circuits the current input vector defines the values of every node in the circuit, in sequential circuits we have memory elements that make the logic functions depend on the previous state of the circuit. As a consequence, there exists a high degree of correlation between the logic values for consecutive clock cycles. In Section 4.1 we present methods to handle pipelined circuits and in Sections 4.2 and 4.3 methods for Finite State Machines (FSMs).

A different kind of sequential correlation is caused by specific input vector sequences. In this case not only do we have significant correlation between clock cycles (temporal correlation), but also correlation between the input signals for a given clock cycle (spatial correlation).

We present a technique to effectively compute the switching activity of a logic circuit under a user-specified input sequence in Section 4.5.

4.1 Pipelines

Many sequential circuits, such as pipelines, can be acyclic. They correspond to blocks of combinational logic separated by flip-flops. An exam-

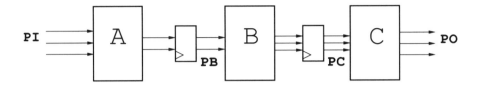

Figure 4.1: A k-pipeline.

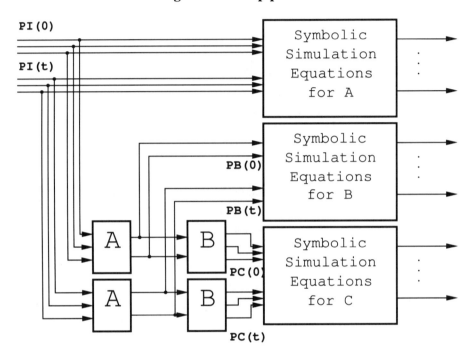

Figure 4.2: Taking k levels of correlation into account.

ple of a 2-stage pipeline, an acyclic sequential circuit, is given in Figure 4.1. *PI* corresponds to the primary inputs to the circuit, *PO* the primary outputs, and *PB* and *PC* the present state lines that are inputs to blocks *B* and *C*, respectively.

It is possible to estimate the power dissipated by acyclic circuits that are k-pipelines, i.e., those that have exactly k flip-flops on each path from primary inputs to primary outputs, without making any assumptions about the probabilities of the present state lines. This is because such circuits are k-definite [4, p. 513], their state and outputs are a function of primary inputs that occurred at most k clock cycles ago.

Consider the circuit of Figure 4.2. The symbolic simulation equa-

4. POWER ESTIMATION FOR SEQUENTIAL CIRCUITS

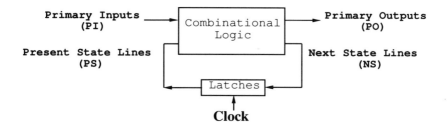

Figure 4.3: A synchronous sequential circuit.

tions corresponding to the switching activities of the logic gates in blocks A, B and C are assumed to have been computed using the method described in Chapter 3. The symbolic simulation equations for block A receive inputs from $PI(0)$ and $PI(t)$, since block A receives inputs from PI alone. The symbolic simulation equations for block B receive inputs from $PB(0)$ and $PB(t)$. To model the relationship between PB and PI, we generate $PB(0)$ from $PI(0)$ and the $PB(t)$ from $PI(t)$. Similarly, the symbolic simulation equations for block C receive inputs from the $PC(0)$ and $PC(t)$ and to model the relationship between PC and PI we generate $PC(0)$ from $PI(0)$ and the $PC(t)$ from $PI(t)$.

In the general case, the symbolic simulation equations corresponding to a combinational logic block in stage l of the pipeline will receive inputs from the cascade of the $l - 1$ previous stages, with inputs $PI(0)$ and $PI(t)$. For a correctly designed pipeline, this models the inputs this logic block will observe l clock cycles later.

The decomposition of Figure 4.2 implies that the gate output switching activity can be determined given only the vector pair $\langle PI(0), PI(t) \rangle$ for the primary inputs. Therefore, to compute gate output transition probabilities, we only require the transition probabilities for the primary inputs. The use of the logic in the previous pipeline stages generates Boolean equations which model the relationship between the state of the circuit and the previously applied input vectors.

4.2 Finite State Machines: Exact Method

In general, sequential circuits are cyclic. A generic sequential circuit is shown in Figure 4.3. Power estimation for these circuits is significantly more complicated.

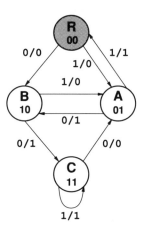

Figure 4.4: Example state transition graph.

We still have the issue of correlation between consecutive clock cycles: the present state lines for the next clock cycle are completely determined by the primary inputs and present state lines of the previous clock cycle. Since a new primary input vector is applied always at the beginning of each clock cycle, correlation between consecutive clock cycles is equivalent to correlation between two input vectors to the combinational logic block.

Further, the probability of the present state lines depends on the probability of the circuit being in any of its possible states. Given a circuit with K flip-flops, there are 2^K possible states. The probability of the circuit being in each state is, in general, not uniform.

As an example, consider a sequential circuit with the State Transition Graph of Figure 4.4 and implemented as in Figure 4.3. Assuming that the circuit was in state **R** at time 0, and that at each clock cycle random inputs are applied, at time ∞ (i.e., steady state) the probabilities of the circuit being in state **R**, **A**, **B**, **C** are $\frac{1}{6}, \frac{1}{3}, \frac{1}{4}$ and $\frac{1}{4}$, respectively. These *state probabilities* have to be taken into account during switching activity estimation of the combinational logic part of the circuit.

4.2.1 Modeling Temporal Correlation

To model the correlation between the two input vectors corresponding to consecutive clock cycles, we append the *next state logic block* to the symbolic network generated for the combinational logic block using the tech-

4. POWER ESTIMATION FOR SEQUENTIAL CIRCUITS

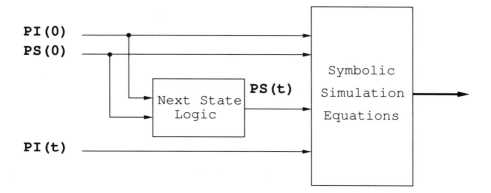

Figure 4.5: Generating temporal correlation of present state lines.

niques described in Chapter 3. The next state logic block is the part of the combinational logic block that computes the next state lines. This augmentation is summarized in Figure 4.5.

The symbolic network has two sets of inputs, namely $PI(0)$ and $PI(t)$ for the primary inputs and $PS(0)$ and $PS(t)$ for the present state lines. However, given $PI(0)$ and $PS(0)$, $PS(t)$ is uniquely determined by the functionality of the combinational logic. This is accomplished by the introduction of the next state logic as shown in Figure 4.5.

The configuration of Figure 4.5 implies that the gate output switching activity can be determined given the vector pair $\langle PI(0), PI(t) \rangle$ for the primary inputs, but only $PS(0)$ for the state lines. Therefore, to compute gate output transition probabilities, we require the transition probabilities for the primary input lines, and the static probabilities for the present state lines.

This configuration was originally proposed in [2].

4.2.2 State Probability Computation

The static probabilities for the present state lines marked $PS(0)$ in Figure 4.5 are spatially correlated. We therefore require knowledge of the *present state probabilities* as opposed to present state line probabilities in order to exactly calculate the switching activity in the sequential circuit. The state probabilities are dependent on the connectivity of the State Transition Graph (STG) of the circuit.

For each state s_i, $1 \leq i \leq K$, in the STG, we associate a variable $prob(s_i)$ corresponding to the steady-state probability of the circuit being in state s_i at $t = \infty$. For each edge e in the STG, we have $e.Current$ signifying the state that the edge fans out from, $e.Next$ signifying the state that the edge fans in to, and $e.Input$ signifying the primary input combination corresponding to the edge. Given static probabilities for the primary inputs[1] to the circuit, we can compute $prob(e.Input)$, the probability of the combination $e.Input$ occurring. We can compute the probability of traversing edge e, $prob(e)$, using

$$prob(e) = prob(e.Current) \times prob(e.Input). \tag{4.1}$$

For each state s_i we can write an equation representing the probability that the machine enters state s_i

$$prob(s_i) = \sum_{\forall e:\ e.Next=s_i} prob(e). \tag{4.2}$$

Given K states, we obtain K equations out of which any one equation can be derived from the remaining $K-1$ equations. We have a final equation

$$\sum_{i=1}^{K} prob(s_i) = 1. \tag{4.3}$$

This linear set of K equations can be solved to obtain the different $prob(s_i)$'s.

This system of equations is known as the Chapman-Kolmogorov equations for a discrete-time discrete-transition Markov process. Indeed, if the Markov process satisfies the conditions that it has a finite number of states, its essential states form a single-chain and it contains no periodic-states, then the above system of equations will have a unique solution [10, pp. 635-654].

For example, for the STG of Figure 4.4 we will obtain the following equations, assuming a probability of 0.5 for the primary input being a 1,

$$prob(\mathbf{R}) = 0.5 \times prob(\mathbf{A}).$$
$$prob(\mathbf{A}) = 0.5 \times prob(\mathbf{R}) + 0.5 \times prob(\mathbf{B}) + 0.5 \times prob(\mathbf{C}).$$
$$prob(\mathbf{B}) = 0.5 \times prob(\mathbf{R}) + 0.5 \times prob(\mathbf{A}).$$

The final equation is

$$prob(\mathbf{R}) + prob(\mathbf{A}) + prob(\mathbf{B}) + prob(\mathbf{C}) = 1.$$

Solving this linear system of equations results in the state probabilities, $prob(\mathbf{R}) = \frac{1}{6}$, $prob(\mathbf{A}) = \frac{1}{3}$, $prob(\mathbf{B}) = \frac{1}{4}$ and $prob(\mathbf{C}) = \frac{1}{4}$.

---

[1] Static probabilities can be computed from specified transition probabilities as given by Equation 2.4.

4. POWER ESTIMATION FOR SEQUENTIAL CIRCUITS

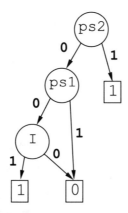

Figure 4.6: BDD for $ps_1 = ps_2 \vee (I \wedge \overline{ps_1})$.

4.2.3 Power Estimation given State Probabilities

Each state corresponds to some binary code. Thus the probabilities computed from the Chapman-Kolmogorov equations are probabilities for each combination of present state lines. The signal probability calculation procedure (described in Section 2.2.2) for the augmented symbolic network of Figure 4.5 has to appropriately weight these combinations according to the given probabilities.

In the case of the BDD-based method, we can still compute signal probabilities taking into account state probabilities with a linear-time traversal. We constrain the ordering of the BDD variables such that all present state lines are necessarily on top. We cannot multiply the probabilities of individual present state lines as we traverse the BDD. We have to save the combination of present state lines encountered during the traversal of each path. At the end of each path, we examine which present state codes are included in the combination of present state lines, add the probabilities computed for these states and multiply this value with the probability obtained from the primary inputs.

For instance, the BDD corresponding to the Boolean function that generates the least significant state line (ps_1) for the FSM of Figure 4.4 is represented in Figure 4.6.

Starting at the left-most $\boxed{1}$, the traversal process will first see the primary input I, then $\overline{ps_1}$ and finish with $\overline{ps_2}$. Therefore the only present state included in this combination of present state lines is 00. The probability corresponding to this term is $prob(I) \times prob(00)$.

ps_2 is the only variable in the path for the right-most $\boxed{1}$. The present states included in ps_2 are 10 and 11, thus the probability for this term is $prob(10) + prob(11)$.

Then, the probability of ps_1 computed using present state probabilities is given by

$$\begin{aligned} prob(ps_1) &= prob(I) \times prob(00) + prob(10) + prob(11) \\ &= prob(I) \times prob(\mathbf{R}) + prob(\mathbf{B}) + prob(\mathbf{C}). \end{aligned}$$

4.3 Finite State Machines: Approximate Method

The Chapman-Kolmogorov system of equations (Equations 4.2 and 4.3) requires the explicit enumeration of all the states in the circuit and this can be very costly. If we have N registers in the circuit the number of possible states is $K = 2^N$. Therefore the exact method is only applicable to small sized circuits, typically with no more than $N = 20$ registers. However, in [3] the authors report solving the Chapman-Kolmogorov system of equations for some large Finite State Machines using Algebraic Decision Diagrams [1].

We describe an approximate method that computes the probabilities of the state lines directly [14]. This way we need only compute N values instead of 2^N. The approximation error comes from the fact that we ignore the correlation between the state lines.

Recently a simulation-based technique to compute state line probabilities has been presented [8]. N logic simulations of the sequential circuit are done starting at some initial state S_0 and the value of each state line is checked at time k. N is determined from the confidence level α and allowed percentage error ϵ. k is the number of cycles the circuit has to go through in order to be considered in steady state. In steady state, the probabilities of the state lines are independent from the initial state, thus N parallel simulations are done starting from some other state S_1. k is determined as the time at which the line probabilities obtained from starting at state S_0 and from S_1 are within ϵ.

4.3.1 Basis for the Approximation

Consider a machine with two registers whose states are 00, 01, 10 and 11 and have state probabilities $prob(00) = \frac{1}{6}, prob(01) = \frac{1}{3}, prob(10) = \frac{1}{4}$

4. POWER ESTIMATION FOR SEQUENTIAL CIRCUITS

and $prob(11) = \frac{1}{4}$. We can calculate the present state line probabilities as shown below, where ps_1 and ps_2 are the first and second present state lines respectively.

$$\begin{aligned}
prob(ps_1 = 0) &= prob(00) + prob(10) = \tfrac{1}{6} + \tfrac{1}{4} = \tfrac{5}{12} \\
prob(ps_1 = 1) &= prob(01) + prob(11) = \tfrac{1}{3} + \tfrac{1}{4} = \tfrac{7}{12} \\
prob(ps_2 = 0) &= prob(00) + prob(01) = \tfrac{1}{6} + \tfrac{1}{3} = \tfrac{1}{2} \\
prob(ps_2 = 1) &= prob(10) + prob(11) = \tfrac{1}{4} + \tfrac{1}{4} = \tfrac{1}{2}
\end{aligned} \quad (4.4)$$

Because ps_1 and ps_2 are correlated, $prob(ps_1 = 0) \times prob(ps_2 = 0) = \frac{5}{24}$ is not equal to $prob(00) = \frac{1}{6}$.

We carried out the following experiment on 52 sequential circuit benchmark examples for which the exact state probabilities could be calculated. These benchmarks included finite state machine controllers, datapaths as well as pipelines. First, the power dissipation of the circuit was calculated using the exact state probabilities as described in Section 4.2.2. Next, given the exact state probabilities, the line probabilities were determined as exemplified in Equation 4.4. Using the topology of Figure 4.5 and the computed present state line probabilities for the *PS* lines, approximate power estimates were calculated for each circuit. The average error (caused by ignoring the correlation between the present state lines) in the power dissipation measures obtained using the line probability approximation over all the circuits was only 2.8%. The maximum error for any one example was 7.3%. Assuming uniform line probabilities of 0.5 as in [2] results in significant errors of over 40% for some examples.

The above experiment leads us to conclude that if accurate line probabilities can be determined then using line probabilities rather than state probabilities is a viable alternative.

4.3.2 Computing Present State Line Probabilities

The approximation framework is based on solving a non-linear system of equations to compute the state line probabilities. This system of equations is given by the combinational logic implementing the next state function of the sequential circuit. The non-linear formulation was developed independently in [7] and [15] and were combined in [14].

Consider the set of functions below corresponding to the next state lines.

$$ns_1 = f_1(i_1, i_2, \cdots, i_M, ps_1, ps_2, \cdots, ps_N)$$
$$ns_2 = f_2(i_1, i_2, \cdots, i_M, ps_1, ps_2, \cdots, ps_N)$$
$$\vdots$$
$$ns_N = f_N(i_1, i_2, \cdots, i_M, ps_1, ps_2, \cdots, ps_N)$$

We can write:

$$prob(ns_1) = prob(f_1(i_1, i_2, \cdots, i_M, ps_1, ps_2, \cdots, ps_N))$$
$$prob(ns_2) = prob(f_2(i_1, i_2, \cdots, i_M, ps_1, ps_2, \cdots, ps_N))$$
$$\vdots$$
$$prob(ns_N) = prob(f_N(i_1, i_2, \cdots, i_M, ps_1, ps_2, \cdots, ps_N))$$

where $prob(ns_i)$ corresponds to the probability that ns_i is a 1, and $prob(f_i(i_1, \cdots, i_M, ps_1, \cdots, ps_N))$ corresponds to the probability that $f_i(i_1, \cdots, i_M, ps_1, \cdots, ps_N)$ is a 1, which is of course dependent on the $prob(ps_j)$ and the $prob(i_k)$.

We are interested in the steady state probabilities of the present and next state lines implying that:

$$prob(ps_i) = prob(ns_i) = p_i \quad 1 \leq i \leq N$$

The set of equations given the values of $prob(i_k)$ becomes:

$$\begin{aligned} p_1 &= g_1(p_1, p_2, \cdots, p_N) \\ p_2 &= g_2(p_1, p_2, \cdots, p_N) \\ &\vdots \\ p_N &= g_N(p_1, p_2, \cdots, p_N) \end{aligned} \quad (4.5)$$

where the g_i's are non-linear functions of the p_i's. We will denote the above equations as $P = G(P)$.

In general the Boolean function f_i can be written as a list of minterms over the i_k and ps_j and the corresponding g_i function can be easily derived. For example, given

$$f_1 = i_1 \wedge ps_1 \wedge \overline{ps_2} \vee i_1 \wedge \overline{ps_1} \wedge ps_2 \quad (4.6)$$

and $prob(i_1) = 0.5$, we have

$$g_1 = 0.5 \cdot (p_1 \cdot (1 - p_2) + (1 - p_1) \cdot p_2) \quad (4.7)$$

We can solve the equation set $P = G(P)$ to obtain the present state line probabilities. The uniqueness or the existence of the solution is not guaranteed for an arbitrary system of non-linear equations. However, since in our application we have a correspondence between the non-linear system of equations and the State Transition Graph of the sequential circuit there will exist at least one solution to the non-linear system.

Obtaining a solution for the given non-linear system of equations requires the use of iterative techniques such as the Picard-Peano or Newton-Raphson methods.

4.3.3 Picard-Peano Method

The Picard-Peano method can be used to find a fixed point of a system of equations of the form $P = G(P)$, such as Equation 4.5. We start with an initial guess P^0, and iteratively compute $P^{k+1} = G(P^k)$ until convergence is reached. Convergence is deemed to be achieved if $P^{k+1} - P^k$ is sufficiently small.

We apply the methods described in Section 2.2.2 to compute $g_i(p_1, p_2, \cdots, p_N)$, the static probability that $f_i(i_1, i_2, \cdots, i_M, ps_1, ps_2, \cdots, ps_N)$ evaluates to 1 for given $p_j = prob(ps_j)$'s and $prob(i_k)$'s.

The use of the Picard-Peano method to solve the system of equations in Equation 4.5 was first proposed in [15]. The convergence proof given in [15, Theorem 3.3] and in [14, Theorem 7.2] is valid only for the single variable case. We present more general convergence conditions.

Theorem 4.1 [9, p. 120] *If G is contractive in a closed set D_0, i.e., $||G(A) - G(B)|| < ||A - B||, \forall A, B \in D_0$, and $G(D_0) \subseteq D_0$ then the Picard-Peano iteration method converges at least linearly to a unique solution P^*.*

Theorem 4.2 *If we have*
$$\sum_{k=1}^{N} \left| \frac{\partial g_i}{\partial p_k} \right| < 1, \ \forall i$$
then G is contractive on the domain $[0,1]^N$.

Proof — First note that $0 \leq g_i \leq 1, \forall i$, therefore $G(D_0) \subseteq D_0$, where $D_0 = [0,1]^N$.

Using the ∞-norm,

$$||G(A) - G(B)||_\infty < ||A - B||_\infty \Leftrightarrow \max_i |g_i(A) - g_i(B)| < \max_j |A_j - B_j|$$

Let $h(t): \mathbb{R} \to \mathbb{R}^N$ such that $h(t) = At + (1-t)B$, i.e., as t goes from 0 to 1, $h(t)$ maps to the line segment connecting points A and B.

Let us define $F(t) = g_i(h(t))$. Then, using the Mean Value theorem [9, p. 68]

$$g_i(A) - g_i(B) = F(1) - F(0) = \frac{dF}{dt}(\xi)$$

for some $\xi \in [0,1]$, where

$$\frac{dF}{dt} = \nabla F \cdot \frac{dh}{dt}.$$

Since $\frac{dh}{dt} = A - B$,

$$g_i(A) - g_i(B) = \sum_{k=1}^{N} \frac{\partial g_i}{\partial p_k}(A_k - B_k).$$

Then for every i,

$$\frac{|g_i(A) - g_i(B)|}{\max_j |A_j - B_j|} = \frac{\left|\sum_{k=1}^{N} \frac{\partial g_i}{\partial p_k}(A_k - B_k)\right|}{\max_j |A_j - B_j|}$$

$$\leq \sum_{k=1}^{N} \left|\frac{\partial g_i}{\partial p_k} \cdot \frac{(A_k - B_k)}{\max_j |A_j - B_j|}\right|$$

$$\leq \sum_{k=1}^{N} \left|\frac{\partial g_i}{\partial p_k}\right|$$

Therefore

$$\sum_{k=1}^{N} \left|\frac{\partial g_i}{\partial p_k}\right| < 1 \Rightarrow |g_i(A) - g_i(B)| < \max_j |A_j - B_j|$$

∎

Theorem 4.3 *If each g_i is a function of only p_j's with $j \leq i$ (i.e., the Jacobian of G is lower triangular), and each next state line is a nontrivial logic function of at least two present state lines, then the Picard-Peano iteration method converges at least linearly to a unique solution P^* on the domain $(0,1)$.*

4. POWER ESTIMATION FOR SEQUENTIAL CIRCUITS

Proof—Choose any p_j. In order to perform the evaluation of $\frac{\partial g_i}{\partial p_j}$ we cofactor f_i with respect to ps_j (Shannon expansion).

$$f_i = ps_j \wedge f_{i\,ps_j} \vee \overline{ps_j} \wedge f_{i\,\overline{ps_j}}$$

$f_{i\,ps_j}$ and $f_{i\,\overline{ps_j}}$ are the cofactors of f_i with respect to ps_j, and are Boolean functions independent of ps_j. We can write:

$$g_i = p_j \cdot prob(f_{i\,ps_j}) + (1 - p_j) \cdot prob(f_{i\,\overline{ps_j}})$$

Differentiating with respect to p_j gives:

$$\frac{\partial g_i}{\partial p_j} = prob(f_{i\,ps_j}) - prob(f_{i\,\overline{ps_j}}) \qquad (4.8)$$

Since we are considering the domain $(0,1)$, which is not inclusive of 0 and 1, and the ns_i's are nontrivial Boolean functions of at least two present state lines for every i, this partial differential is strictly less than one, because we are guaranteed that $prob(f_{i\,ps_j}) > 0$ and $prob(f_{i\,\overline{ps_j}}) > 0$.

Since $p_1 = g_1(p_1)$ and $\left|\frac{\partial g_1}{\partial p_1}\right| < 1$, g_1 is contractive and Theorem 4.1 can be used for the single variable case to guarantee that the Picard-Peano iterations on g_1 will converge at least linearly to the unique solution p_1^*.

We can now substitute p_1^* in g_2, making g_2 a function of a single variable p_2. The observations of the previous paragraph apply for g_2, thus we obtain p_2^*.

We repeat this process for the remaining g_i's to compute $P^* = (p_1^*, p_2^*, \cdots, p_N^*)$. ∎

We have a somewhat restrictive condition for Theorem 4.2 which is probably not met for a generic logic circuit. However, the conditions for Theorem 4.3 are met for many datapath circuits, where the least significant bit is only a function of the least significant bit of the input and a carry is generated which is only a function of lower order bits.

We should stress that Theorems 4.2 and 4.3 are sufficient, and not necessary, conditions for convergence. In practice, we have observed that for most of the circuits, even those not in the conditions of these theorems, Picard-Peano rapidly converges to a solution.

4.3.4 Newton-Raphson Method

The Newton-Raphson method can be used to solve a non-linear system of equations of the form $Y(P) = 0$. We rewrite the system of Equation 4.5

to be in the form

$$\begin{aligned} y_1 &= p_1 - g_1(p_1, p_2, \cdots, p_N) = 0 \\ y_2 &= p_2 - g_2(p_1, p_2, \cdots, p_N) = 0 \\ &\vdots \\ y_N &= p_N - g_N(p_1, p_2, \cdots, p_N) = 0. \end{aligned} \quad (4.9)$$

The advantage of the Newton-Raphson method is that it is more robust in terms of convergence requirements and its rate of convergence is quadratic instead of linear. However, each iteration is more computationally expensive than the Picard-Peano method. The use of the Newton-Raphson method to solve the system of equations above was first proposed in [7].

Given $Y(P) = 0$ and a column matrix corresponding to an initial guess P^0, we can write the kth Newton iteration as the linear system shown below

$$J(P^k) \times P^{k+1} = J(P^k) \times P^k - Y(P^k), \quad (4.10)$$

where J is the $N \times N$ Jacobian matrix of the system of equations Y. Each entry in J corresponds to a $\frac{\partial y_i}{\partial p_j}$ evaluated at P^k. The P^{k+1} corresponds to the variables in the linearized system and after solving the system P^{k+1} is used as the next guess. Convergence is deemed to be achieved if each entry in $Y(P^k)$ is sufficiently small.

Again, the methods of Section 2.2.2 are used to evaluate

$$g_i(p_1, p_2, \cdots, p_N) = prob(f_i(i_1, i_2, \cdots, i_M, ps_1, ps_2, \cdots, ps_N)),$$

for given $p_j = prob(ps_j)$ and $prob(i_k)$. The $Y(P^k)$ of Equation 4.10 can easily be evaluated using the p_j^k values and Equation 4.9.

We need to also evaluate $J(P^k)$. As mentioned earlier, each entry of J corresponds to $\frac{\partial y_i}{\partial p_j}$ evaluated at P^k. If $i \neq j$, then $\frac{\partial y_i}{\partial p_j}$ equals $-\frac{\partial g_i}{\partial p_j}$, and $\frac{\partial y_i}{\partial p_i}$ equals equals $1 - \frac{\partial g_i}{\partial p_i}$.

In order to perform the evaluation of $\frac{\partial y_i}{\partial p_j}$ we use the result obtained in Theorem 4.3 (Equation 4.8)

$$\frac{\partial y_i}{\partial p_j} = prob(f_i{}_{\overline{ps_j}}) - prob(f_i{}_{ps_j}) \quad (4.11)$$

We can evaluate $prob(f_i{}_{ps_j})$ and $prob(f_i{}_{\overline{ps_j}})$ for a given P^k again using the methods of Section 2.2.2.

4. POWER ESTIMATION FOR SEQUENTIAL CIRCUITS

As an example, consider the function given in Equation 4.6:

$$f_1 = i_1 \wedge ps_1 \wedge \overline{ps_2} \vee i_1 \wedge \overline{ps_1} \wedge ps_2$$
$$\frac{\partial g_1}{\partial p_1} = prob(i_1 \wedge \overline{ps_2}) - prob(i_1 \wedge ps_2)$$
$$= 0.5 \cdot (1 - p_2) - 0.5 \cdot p_2$$
$$= 0.5 - p_2$$

which is exactly what we obtain by differentiating Equation 4.7 with respect to p_1.

Theorem 4.4 [9, p. 412] *The Newton iterates:*

$$P^{k+1} = P^k - J(P^k)^{-1} Y(P^k), k = 0, 1, \ldots$$

are well-defined and converge to a solution P^* of $Y(P) = 0$ in the domain D_0, if the following conditions are satisfied:

1. *Y is F-differentiable.*
2. *$\|J(A) - J(B)\| \leq \gamma \|A - B\|, \forall A, B \in D_0$*
3. *There exists some $P^0 \in D_0$ such that $\left\|J(P^0)^{-1} Y(P^0)\right\| \leq \eta$, $\left\|J(P^0)^{-1}\right\| \leq \beta$ and $\alpha = \frac{1}{2}\beta\gamma\eta < 1$.*

Condition 1 of the theorem is satisfied in our application because the y_i functions are continuous and differentiable.

To show that Condition 2 is satisfied, we need to prove that the parameter γ is finite for all A,B in the domain. In our case the domain is defined by $0 \leq a_i, b_i \leq 1$ for $1 \leq i \leq N$, where N is the dimension of Y.

Theorem 4.5 *If Y is given by Equation 4.9, then Condition 2 of Theorem 4.4 is satisfied for $\gamma = 2N$.*

Proof— We will use the 1-norm to show that

$$\|J(A) - J(B)\|_1 \leq \gamma \|A - B\|_1, \forall A,B \in D_0$$

is satisfied for $\gamma = 2N$.

Let $h(t) : \mathbb{R} \to \mathbb{R}^N$ such that $h(t) = At + (1 - t)B$, i.e., as t goes from 0 to 1, $h(t)$ maps to the line segment connecting points A and B.

Recall that each entry in matrix J is given by $\frac{\partial y_i}{\partial p_j}$. Let us define $F(t) = \frac{\partial y_i}{\partial p_j}(h(t))$. Again using the Mean Value theorem [9, p. 68]

$$\frac{\partial y_i}{\partial p_j}(A) - \frac{\partial y_i}{\partial p_j}(B) = F(1) - F(0) = \frac{dF}{dt}(\xi)$$

for some $\xi \in [0,1]$, where

$$\frac{dF}{dt} = \nabla F \cdot \frac{dh}{dt} = \sum_{k=1}^{N} \frac{\partial^2 y_i}{\partial p_j \partial p_k} \frac{dh_k}{dt}$$

In order to perform the evaluation of $\frac{\partial y_i}{\partial p_j}$ we use Equation 4.11:

$$\frac{\partial y_i}{\partial p_j} = \text{prob}(f_i \, \overline{ps_j}) - \text{prob}(f_i \, ps_j)$$

Differentiating with respect to p_k we have:

$$\frac{\partial^2 y_i}{\partial p_j \partial p_k} = \text{prob}(f_i \, \overline{ps_j} ps_k) - \text{prob}(f_i \, \overline{ps_j}\,\overline{ps_k}) - \text{prob}(f_i \, ps_j ps_k) + \text{prob}(f_i \, ps_j \overline{ps_k})$$

Given that the probabilities are between 0 and 1, we have:

$$\left| \frac{\partial^2 y_i}{\partial p_j \partial p_k} \right| \leq 2$$

Then

$$\left| \frac{\partial y_i}{\partial p_j}(A) - \frac{\partial y_i}{\partial p_j}(B) \right| = \left| \sum_{k=1}^{N} \frac{\partial^2 y_i}{\partial p_j \partial p_k} \frac{dh_k}{dt} \right|$$

$$\leq \sum_{k=1}^{N} \left| \frac{\partial^2 y_i}{\partial p_j \partial p_k} \frac{dh_k}{dt} \right|$$

$$\leq 2 \sum_{k=1}^{N} \left| \frac{dh_k}{dt} \right|$$

On the other hand, $\frac{dh}{dt} = A - B$. From the definition of 1-norm,

$$\|A - B\|_1 = \sum_{k=1}^{N} |a_k - b_k| = \sum_{k=1}^{N} \left| \frac{dh_k}{dt} \right|,$$

therefore
$$\left|\frac{\partial y_i}{\partial p_j}(A) - \frac{\partial y_i}{\partial p_j}(B)\right| \leq 2 \cdot ||A - B||_1$$

The 1-norm of a matrix M is defined as
$$||M||_1 = \max_{\hat{u} \neq 0} \frac{||M \cdot \hat{u}||_1}{||\hat{u}||_1} = \max_k \sum_{i=1}^{N} |m_{ik}|.$$

Since each entry in the matrix $J(A) - J(B)$ is bounded by $2 \cdot ||A - B||_1$,
$$\begin{aligned}
||J(A) - J(B)||_1 &= \max_k \sum_{i=1}^{N} \left|\frac{\partial y_i}{\partial p_k}(A) - \frac{\partial y_i}{\partial p_k}(B)\right| \\
&\leq \max_k \sum_{i=1}^{N} |2 \cdot ||A - B||_1| \\
&= 2 \cdot N \cdot ||A - B||_1.
\end{aligned}$$

Therefore, $\gamma = 2N$. ∎

If we are in the conditions of Theorem 4.3, we can show some results about the norm of the inverse of the Jacobian.

Theorem 4.6 *If*
$$\sum_{k=1}^{N} \left|\frac{\partial g_i}{\partial p_k}\right| < 1, \ \forall i$$
then $\left||J(P^0)^{-1}\right||$ is finite, for all $P^0 \in D_0$.

Proof — The entries of row i in J are $\frac{\partial y_i}{\partial p_i} = 1 - \frac{\partial g_i}{\partial p_i}$ and $\frac{\partial y_i}{\partial p_j} = -\frac{\partial g_i}{\partial p_j}, \forall j \neq i$. If $\sum_{j=1}^{N} \left|\frac{\partial g_i}{\partial p_j}\right| < 1$, then $J(P^0)$ is diagonally dominant and thus is invertible [9, p. 48]. This means that $||J(P^0)|| \neq 0$, therefore $||J^{-1}(P^0)||$ is bounded by some positive β. ∎

Condition 3 in Theorem 4.4 is a constraint on the initial guess for the Newton iteration, and this initial guess can be picked appropriately, provided γ is finite. Essentially, we have to choose P^0 such that $\beta\eta \leq \frac{1}{N}$.

Note that if the condition for Theorem 4.6 is met, not only do we have a bound for β of Theorem 4.4, but also a bound for η. Since $y_i \leq 1$,

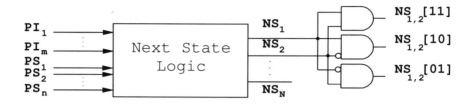

Figure 4.7: An *m*-expanded network with $m = 2$.

$\|Y(P^0)\|_1 \leq N$. Then

$$\begin{aligned} \|J(P^0)^{-1} Y(P^0)\|_1 &= \|J(P^0)^{-1}\|_1 \cdot \|Y(P^0)\|_1 \\ &\leq N \cdot \|J(P^0)^{-1}\|_1 \\ &\leq N \cdot \beta = \eta. \end{aligned}$$

4.3.5 Improving Accuracy using m-Expanded Networks

The above formulation does not capture the correlation between the state line probabilities. While the state line probabilities obtained using the above method will result in switching activity estimates that are close to the exact method, it is worthwhile to explore ways to increasing the accuracy.

In this section we describe a method models the correlation between *m*-tuples of present state lines. The method is pictorially illustrated in Figure 4.7 for $m = 2$.

The number of equations in the case of $m = 2$ is $\frac{3N}{2}$. We have:

$$\begin{aligned} ns_{i,i+1}[11] &= ns_i \wedge ns_{i+1} = f_i \wedge f_{i+1} \\ ns_{i,i+1}[10] &= ns_i \wedge \overline{ns_{i+1}} = f_i \wedge \overline{f_{i+1}} \\ ns_{i,i+1}[01] &= \overline{ns_i} \wedge ns_{i+1} = \overline{f_i} \wedge f_{i+1} \end{aligned}$$

We have to solve for $prob(ns_{i,i+1}[11])$, $prob(ns_{i,i+1}[10])$, and $prob(ns_{i,i+1}[01])$ (rather than $prob(ns_i)$ and $prob(ns_{i+1})$ as in the case of $m = 1$). We use:

$$\begin{aligned} prob(ps_i \wedge ps_{i+1}) &= prob(ns_{i,i+1}[11]) \\ prob(ps_i \wedge \overline{ps_{i+1}}) &= prob(ns_{i,i+1}[10]) \\ prob(\overline{ps_i} \wedge ps_{i+1}) &= prob(ns_{i,i+1}[01]) \end{aligned}$$

in the evaluation of the $prob(f_i)$'s.

4. POWER ESTIMATION FOR SEQUENTIAL CIRCUITS

For $m = 3$ the number of equations is $\frac{7N}{3}$. For general m, the number of combinations we have for each m-tuple is $2^m - 1$ and the number of m-tuples is N/m, therefore the total number of equations in the non-linear system is

$$\frac{(2^m - 1)N}{m}$$

$m = 1$ corresponds to the approximate method presented in the previous sections. When $m = N$, the number of equations will become $2^N - 1$ and the method degenerates to the Chapman-Kolmogorov method.

The choice of the m-tuples of present and next state lines is made by grouping next state lines that have the maximal amount of shared logic into each m-tuple. Note that the accuracy of line probability estimation will depend on the choice of the m-tuples.

The signal probability evaluation during the iterations to solve the non-linear system of equations has to use the probability for each combination of each m-tuple. This is done in the same way as for the exact power estimation method of Section 4.2.3 where state probabilities are required.

To estimate switching activity given m-tuple present state line probabilities, the topology of Figure 4.5 is used as before. Again the signal probability has to be computed using the probabilities of the m-tuples.

4.3.6 Improving Accuracy using k-Unrolled Networks

The topology of Figure 4.5 was proposed as a means of taking into account the correlation between the applied input vector pair when computing the transition probabilities. This method takes one cycle of correlation into account. It is possible to take multiple cycles of correlation into account by prepending the symbolic simulation equations with the k-unrolled network. This is illustrated in Figure 4.8. Instead of connecting the next state logic network to the symbolic simulation equations, we unroll the next state logic network k times and connect the next state lines of the kth stage of the unrolled network, the next state lines of the $k - 1^{th}$ stage, and the primary inputs of the $k - 1^{th}$ stage to the symbolic simulation equations.

Each next state logic level introduces correlation between present state lines PS^j, thus making the switching activity computed by the symbolic network be closer to the exact method. In the limiting case, when $k \to \infty$, the k-unrolled network will compute the exact switching activity, independently of the value used for PS^0.

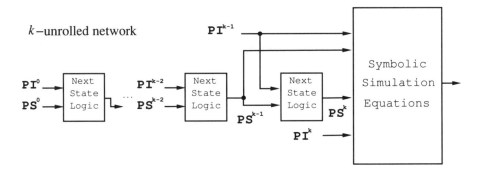

Figure 4.8: Calculation of signal and transition probabilities by network unrolling.

Recall that the error is introduced by ignoring the correlation between the present state lines. For the exact method of Section 4.2 where state probabilities are used, one stage of the next state logic suffices to obtain the exact switching activity.

4.3.7 Redundant State Lines

In this section we make some observations on how the approximate methods behave in the presence of redundant state lines. Let us consider a case study. The STG of Figure 4.9(a) describes a system with two outputs, the first output $O1$ is 1 when the input I is 0 at even clock cycles and the second output $O2$ is 1 when the input I is 1 at odd clock cycles. A minimum logic implementation of this circuit is depicted in Figure 4.9(b).

Assume that for some reason the designer prefers to choose the implementation of Figure 4.10(a). This circuit has the same input/output behavior as the circuit of Figure 4.9(b), but we have a redundant present state line.

First note that the exact method will compute the same probabilities for the states as the STG for the circuit stays the same.

For the approximate method, we get the equations:

$$ns_1 = \overline{ps_1}$$
$$ns_2 = \overline{ps_1}$$

which rewritten in terms of probabilities become

$$p_1 = 1 - p_1$$
$$p_2 = 1 - p_1$$

4. POWER ESTIMATION FOR SEQUENTIAL CIRCUITS

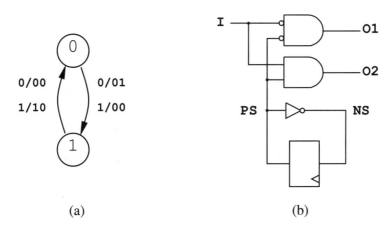

Figure 4.9: **Example circuit: (a) State transition graph; (b) Logic circuit.**

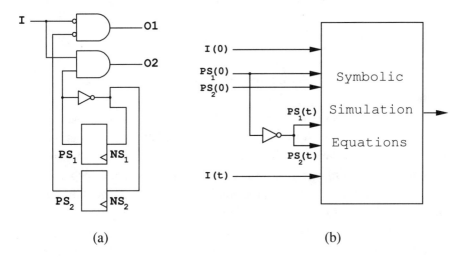

Figure 4.10: **(a) Circuit with a redundant state line; (b) 1-unrolled symbolic network.**

The conditions for Theorem 4.2 are not verified since $\frac{\partial g_1}{\partial p_1} = \frac{\partial g_2}{\partial p_1} = -1$. Also Theorem 4.3 is not applicable as ns_1 and ns_2 are a function of a single present state line. In fact, if we start with $p_1 \neq 0.5$, the Picard-Peano method will oscillate. However, this is not related to the redundant present state line, the oscillation problem remains the same for the implementation of Figure 4.9(b).

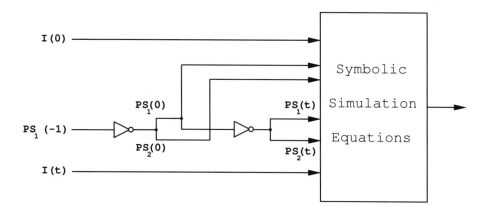

Figure 4.11: Symbolic network with $k = 2$.

The Jacobian J for this system of equations is

$$J = \begin{bmatrix} 2 & 0 \\ 1 & 1 \end{bmatrix}$$

so J^{-1} is well defined and the Newton-Raphson will converge to the right solution.

The symbolic simulation network with one stage of the next state logic ($k = 1$) is shown in Figure 4.10(b) (the next state logic block is just an inverter). If we use the approximate power estimation method on this network, we will be introducing a large error because the correlation between the present state lines is very large. Although using Newton-Raphson we will obtain the same probabilities for the two state lines $PS_1(0)$ and $PS_2(0)$, the symbolic network does not have enough information to indicate that these lines are actually the same. Still, the next state logic introduces this information for $PS_1(t)$ and $PS_2(t)$.

By unrolling the circuit one more time, as shown in Figure 4.11, we introduce the necessary correlation between $PS_1(0)$ and $PS_2(0)$. Thus with $k = 2$ we can achieve the exact solution with the approximate method. However, it is not true that for the general case $k = 2$ suffices.

Similarly, if we use the m-expand method with $m = 2$ the exact solution will be obtained.

4. POWER ESTIMATION FOR SEQUENTIAL CIRCUITS

CKT NAME	GATES	S	FF	UNIFORM PROB.			LINE PROB.			PIPELINE	
				P	ERR	CPU	P	ERR	CPU	P	CPU
mult4	40	1	14	769	24.4	4s	621	0.5	4s	618	4s
mult8	176	1	46	3815	16.4	72s	3290	0.4	76s	3278	3.9m
		2	87	4440	20.6	58s	3691	0.3	75s	3680	4.3m
		3	136	5145	23.8	65s	4166	0.3	90s	4154	4.6m
cla16	150	1	37	1249	4.7	10s	1194	0.0	12s	1193	14s
		2	66	1609	5.9	12s	1521	0.1	16s	1520	15s
		3	102	2050	6.9	13s	1919	0.0	20s	1918	18s
cbp32	353	1	54	2596	4.1	36s	2491	0.1	39s	2494	41s
		2	111	3290	6.7	37s	3069	0.5	48s	3083	44s
		3	162	3930	7.8	40s	3628	0.5	63s	3648	52s

Table 4.1: Comparison of sequential power estimation methods for pipelined circuits.

4.4 Results on Sequential Power Estimation Techniques

In this section we present experimental results that illustrate the following points:

- Purely combinational logic estimates result in significant inaccuracies.

- Assuming uniform probabilities for the present state line probabilities and state probabilities as in [2] can result in significant inaccuracies in power estimates.

- Exact and explicit computation of state probabilities is possible for controller type circuits. However, it is not viable for datapath circuits.

- For acyclic datapath circuits, the method described in Section 4.1 produces exact results and is very efficient since no state or line probabilities need to be computed.

- Computing the present state *line* probabilities using the techniques presented in the previous sections results in 1) accurate switching activity estimates for all internal nodes in the network implementing the sequential machine; 2) accurate, robust and computationally efficient power estimate for the sequential machine.

In Table 4.1, results are presented for pipelined datapath circuits. We present results for a 4- and a 8-bit multipliers and carry-look-ahead and carry-bypass adders, for different number of pipeline stages. For each circuit, we give the number of gates in the combinational logic block (GATES) and the total number of flip-flops (FF) for each number of pipeline stages (S).

In the table, UNIFORM PROB corresponds to the sequential estimation method assuming uniform (0.5) probabilities for the present state lines. The column LINE PROB corresponds to the approximate technique of Section 4.3 and using the Newton-Raphson method to solve the non-linear system of Equation 4.9. These equations correspond to 1-unrolled or 1-expanded networks. PIPELINE corresponds to the power estimation method for acyclic circuits of Section 4.1.

For the power estimates, we used the symbolic simulation method described in Chapter 3. A zero delay model was assumed, however any other delay model could have been used instead. We give the estimated power for each method under P and the CPU time required under CPU. Under ERR we give the percentage difference relative to the exact method, in this case the PIPELINE method. The CPU times in the table correspond to seconds on a SUN SPARC-2. These are the time required to estimate combinational switching activity using BDD-based symbolic simulation plus the time required for the calculation of state/line probabilities.

Assuming a uniform probability of the present state lines (UNIFORM PROB) can yield very large errors. We can see that if the approximate method of Section 4.3 is used (LINE PROB), the power estimates are very close to that obtained by the exact method (PIPELINE).

For the examples presented in Table 4.1, we were only able to run the exact state probability calculation method of Section 4.2 for `mult4`. As expected, we obtained the same power value as with the PIPELINE but it took 4.8 minutes of CPU time. For the other circuits, the corresponding STG is very large. The method of Section 4.1 (PIPELINE) exactly computes the average switching activity for a pipelined circuit, taking into account the correlation between the flip-flops. It requires much less CPU time since no state probabilities have to be computed.

Tables 4.2 and 4.3 present results for several cyclic circuits, the statistics for which are given in Table 4.2. In these tables, UNIFORM PROB and LINE PROB correspond to the same methods as in Table 4.1. COMBINATIONAL corresponds to the purely combinational estimation method of Chapter 3, i.e.,

4. POWER ESTIMATION FOR SEQUENTIAL CIRCUITS

Circuit Name	Gates	FF	Combinational			Uniform Prob.		
			P	ERR	CPU	P	ERR	CPU
cse	132	4	610	58.7	1s	578	50.3	7s
dk16	180	5	1078	3.1	1s	1097	5.0	10s
dfile	119	5	923	32.5	1s	702	0.6	7s
keyb	169	5	750	43.3	1s	725	38.6	12s
mod12	25	4	245	21.7	0s	196	2.7	1s
planet	327	6	1641	2.5	2s	1709	1.5	17s
sand	336	5	1446	33.1	2s	1166	7.2	24s
sreg	9	3	128	1.4	0s	129	0.0	0s
styr	313	5	1395	45.3	2s	1208	25.8	22s
tbk	478	5	1958	24.1	4s	1904	20.7	48s
accum4	45	4	361	3.5	0s	374	0.0	2s
accum8	89	8	721	4.2	1s	753	0.0	7s
accum16	245	16	1521	-	2s	1596	-	234s
count4	19	4	256	20.1	0s	213	0.0	1s
count7	35	7	474	12.2	0s	423	0.0	2s
count8	40	8	560	10.2	0s	508	0.0	3s
s953	418	29	762	76.8	1s	673	56.0	10s
s1196	529	18	2558	-	4s	2538	-	484s
s1238	508	18	2709	-	4s	2688	-	156s
s1423	657	74	6017	-	251s	4734	-	271s
s5378	4212	164	12457	-	74s	12415	-	455s
s13207	11241	669	37842	-	5m	27186	-	11m
s15850	13659	597	40016	-	8m	23850	-	14m
s35932	28269	1728	122131	-	20m	118475	-	36m
s38584	32910	1452	112706	-	24m	85842	-	44m

Table 4.2: Comparison of power estimation methods for cyclic circuits.

no next state logic block is appended to the symbolic network thus there is no correlation between the two input vectors. STATE PROB corresponds to the exact state probability calculation method of Section 4.2.

The first set of circuits corresponds to finite state machine controllers. These circuits typically have the characteristic that the state probabilities are highly non-uniform. Restricting oneself to combinational power dissipation (COMBINATIONAL) or assuming uniform state probabilities (UNIFORM PROB) results in significant errors. However, the line probability method of Section 4.3 produces highly accurate estimates when compared to exact state probability

Circuit Name	Line Prob. P	Line Prob. Err	Line Prob. CPU	State Prob. P	State Prob. CPU
cse	380	1.0	9s	384	11s
dk16	1045	0.0	13s	1045	15s
dfile	701	0.6	8s	697	10s
keyb	518	1.0	14s	523	15s
mod12	199	1.1	1s	201	1s
planet	1686	0.1	24s	1684	28s
sand	1078	0.7	27s	1086	34s
sreg	129	0.0	0s	129	1s
styr	997	3.8	28s	960	30s
tbk	1538	2.4	52s	1577	71s
accum4	374	0.0	2s	374	5s
accum8	753	0.0	8s	753	875s
accum16	1596	-	239s	unable	
count4	213	0.0	1s	213	2s
count7	423	0.0	3s	423	5s
count8	508	0.0	4s	508	8s
s953	439	1.7	12s	431	15s
s1196	2294	-	488s	unable	
s1238	2439	-	151s	unable	
s1423	7087	-	289s	unable	
s5378	6496	-	478s	unable	
s13207	10573	-	338m	unable	
s15850	10534	-	167m	unable	
s35932	62292	-	152m	unable	
s38584	63995	-	922m	unable	

Table 4.3: **Comparison of power estimation methods for cyclic circuits (contd).**

calculation.

The second set of circuits corresponds to datapath circuits, such as counters and accumulators. The exact state probability evaluation method requires huge amounts of CPU time for even the medium-sized circuits, and cannot be applied to the large circuits. For all the circuits that the exact method is viable for, our LINE PROB method produces identical estimates. The UNIFORM PROB method does better for the datapath circuits – in the case of counters for instance, it can be shown that the state probabilities are all uniform,

4. POWER ESTIMATION FOR SEQUENTIAL CIRCUITS

Circuit Name	Combinational Error	Uniform Prob. Error	Line Prob. Error
cse	NA	0.427	0.00788
dk16	NA	0.0782	0.0125
dfile	NA	0.075	0.047
keyb	NA	0.414	0.0133
mod12	NA	0	0.03
planet	NA	0.031	0.09
sand	NA	0.12	0.044
sreg	NA	0	0
styr	NA	0.3138	0.0357
tbk	NA	0.2614	0.026
accum4	NA	0	0
accum8	NA	0	0
accum16	NA	0	0
count4	NA	0	0
count7	NA	0	0
count8	NA	0	0

Table 4.4: Absolute errors in present state line probabilities averaged over all present state lines.

and therefore the UNIFORM PROB method will produce the right estimates. Of course, this assumption is not always valid.

The third set of circuits corresponds to mixed datapath/control circuits from the ISCAS-89 benchmark set. Exact state probability evaluation is not possible for these circuits.

In Table 4.4, present state line probability estimates for the benchmark circuits are presented. The error value provided in each column shows the absolute error (absolute value of the difference between exact and approximate value) of the signal probability values *averaged* over all present state lines in the circuit. The exact values were calculated by the method described in Section 4.2. It is evident from these results that the error for the approximate method of Section 4.3 averaged over all benchmark circuits is well below 0.05.

We present the switching activity errors for the benchmark circuits in Table 4.5. Again, the error value provided in each column represents the absolute error averaged over all internal nodes in the circuit. It can be seen that this error is quite small. These two tables demonstrate that the approximate

Circuit Name	Combinational Error	Uniform Prob. Error	Line Prob. Error
cse	0.402	0.053	0.003
dk16	0.354	0.020	0.010
dfile	0.268	0.019	0.015
keyb	0.363	0.067	0.009
mod12	0.387	0.149	0.156
planet	0.375	0.034	0.034
sand	0.400	0.015	0.010
sreg	0	0	0
styr	0.415	0.058	0.022
tbk	0.423	0.020	0.008
accum4	0.084	0	0
accum8	0.086	0	0
accum16	0.096	0	0
count4	0.169	0	0
count7	0.189	0	0
count8	0.192	0	0

Table 4.5: Absolute errors in switching activity averaged over all circuit lines.

procedure provided in Section 4.3 leads to very accurate estimates for both the present state line probabilities and for the switching activity values for all circuit lines.

Next, we present results comparing the Picard-Peano and Newton-Raphson methods to solve the non-linear equations of Section 4.3. These results are summarized in Table 4.6. The number of iterations required for the Picard-Peano and Newton-Raphson methods are given in Table 4.6 under the appropriate columns, as are the CPU times per iteration and the total CPU time. Newton-Raphson typically takes fewer iterations, but each iteration requires the evaluation of the Jacobian and is more expensive than the Picard iteration. The results obtained by the two methods are identical, since the convergence criterion used was the same.

The convergence criterion allowed a maximum error of 1% in the line probabilities. In this case, the Picard-Peano method outperforms the Newton-Raphson method for virtually all the examples. If the convergence criterion is tightened, e.g., to allow for a maximum error of 0.01%, the Picard-Peano

4. POWER ESTIMATION FOR SEQUENTIAL CIRCUITS

Circuit Name	Picard-Peano			Newton-Raphson		
	#ITER	CPU/ITER	TOTAL CPU	#ITER	CPU/ITER	TOTAL CPU
cse	5	0.1	0.5	3	1	3
dk16	4	0.18	0.7	3	1	3
dfile	5	0.12	0.6	2	1.5	3
keyb	10	0.07	0.7	6	0.33	2
mod12	3	0.03	0.1	2	0.1	0.2
planet	11	0.13	1.4	3	2.33	7
sand	6	0.22	1.3	3	1	3
sreg	1	0.1	0.1	1	0.1	0.1
styr	7	0.2	1.4	3	2	6
tbk	4	0.5	2.0	3	1.33	4
accum4	1	0.1	0.1	1	0.1	0.1
accum8	1	0.3	0.3	1	1	1
accum16	1	1.0	1.0	1	6	6
count4	1	0.1	0.1	1	0.1	0.1
count7	1	0.2	0.2	1	1	1
count8	1	0.2	0.2	1	1	1
s953	30	0.04	1.1	4	0.5	2
s1196	2	1.1	2.2	2	2	4
s1238	2	1.15	2.3	2	2.5	5

Table 4.6: Comparison of Picard-Peano and Newton-Raphson.

method requires substantially more iterations than the Newton-Raphson and in several examples, the Newton-Raphson method outperforms the Picard-Peano method. However, since the error due to ignoring correlation (cf. Section 4.3) is more than 1% it does not make sense to tighten the convergence criterion beyond a 1% allowed error.

In some examples the Picard-Peano method may exhibit oscillatory behavior, and will not converge. In these cases, the strategy we adopt is to use Picard-Peano for several iterations, and if oscillation is detected, the Newton-Raphson method is applied.

In Table 4.7, we present results that indicate the improvement in accuracy in power estimation when k-unrolled or m-expanded networks are used. Results are presented for the finite state machine circuits of Table 4.2 for $k = 1,2,3$ and $m = 1,2,4$ (the initial error for dk16 and sreg benchmarks is 0, thus there is no need to improve the accuracy by using larger values of k and m).

| Circuit | Initial | k-Unrolled Error | | | | m-Expanded Error | | | |
| Name | Error | $k = 2$ | | $k = 3$ | | $m = 2$ | | $m = 4$ | |
		ERR	CPU	ERR	CPU	ERR	CPU	ERR	CPU
cse	1.06	0.33	18	0.02	51	0.42	10	0.00	10
dfile	0.67	0.20	16	0.20	29	0.23	9	0.17	10
keyb	1.02	0.02	44	0.04	53	1.01	14	0.32	14
mod12	1.13	0.85	2	0.30	3	1.13	1	0.00	2
planet	0.11	0.15	40	1.72	45	0.10	25	0.08	25
sand	0.76	0.61	64	0.29	109	0.64	28	0.43	30
styr	3.85	0.16	67	0.41	113	0.58	29	0.52	29
tbk	2.46	1.52	207	0.12	597	2.17	58	0.12	59

Table 4.7: Results of power estimation using k-unrolled and m-expanded networks.

The percentage differences in power from the exact power estimate are given. If $k \rightarrow \infty$ the error will reduce to 0%, however, increasing k when k is small is not guaranteed to reduce the error (e.g., consider styr). The m-expansion-based method behaves more predictably for this set of examples, however, no guarantees can be made regarding the improvement in accuracy on increasing m, except that when m is set to the number of flip-flops in the machine, the method produces the Chapman-Kolmogorov equations, and therefore the exact state probabilities are obtained. The CPU times for power estimation are in seconds on a SUN SPARC-2. These times can be compared with those listed in Table 4.3 under the LINE PROB column as those times correspond to $k = 1$ and $m = 1$.

During the synthesis process, we often want to know the switching activity of individual nodes instead of a single power consumption figure. Table 4.8 presents the percentage error for individual node's switching activity from the exact values as a function of k and m, averaged over all the nodes in the circuit. It is seen that the accuracy of switching activity estimates consistently increases with the value of k and m. For example, the error in switching activity estimates for styr decreases from 13% to 6.3% when k increases from 2 to 3. The power estimates, however, do not necessarily improve by increasing k or m. This phenomenon can be explained as follows. The total power estimate is obtained by summing power consumptions of all nodes in the circuit. The individual power estimates may be under- or over-estimated, yet when they are

Circuit	k-Unrolled Error			m-Expanded Error		
Name	$k=1$	$k=2$	$k=3$	$m=1$	$m=2$	$m=4$
cse	6.79	2.26	0.57	6.79	3.40	0.00
dfile	14.05	5.37	3.10	14.05	4.82	3.56
keyb	7.18	1.68	0.70	7.18	7.09	2.25
mod12	10.24	6.36	5.00	10.24	10.05	0.00
planet	43.08	30.22	28.97	43.08	41.26	35.22
sand	16.65	12.20	11.78	16.65	14.02	9.42
styr	43.51	12.99	6.31	43.51	6.55	5.97
tbk	18.04	4.48	2.95	18.04	15.91	1.88

Table 4.8: Percentage error in switching activity estimates averaged over all nodes in the circuit.

added together, the overall error may become small due to error cancellation. Increasing k improves the accuracy of power estimates for individual nodes, but does not necessarily improve the accuracy of power estimate for the circuit due to the unpredictability of the error cancellation during the summing step.

4.5 Modeling Correlation of Input Sequences

One of the limitations of the approaches of the previous sections is that the input sequences to the sequential circuit are assumed to be uncorrelated. In reality, the inputs come from other sequential circuits, or are application programs. A high degree of correlation could exist in the applied input sequence. This correlation could be temporal, i.e., consecutive vectors could bear some relationship, or could be spatial, i.e., bits within a vector could bear some relationship.

Recently a technique was proposed in [5] that tries to introduce some degree of information about correlation between inputs. This estimation method allows the user to specify pairwise correlation of inputs as *static* (SC) and *transition* (TC) *correlation coefficients*. These are defined as:

$$SC_{ij}^{xy} = \frac{prob(x=i \wedge y=j)}{prob(x=i)prob(y=j)} \qquad TC_{ij,kl}^{xy} = \frac{prob(x_i \rightarrow k \wedge y_j \rightarrow l)}{prob(x_i \rightarrow k)prob(y_j \rightarrow l)}$$

These coefficients are then propagated through the logic circuit and similar coefficients for internal signals are obtained. This results in efficient estimation schemes, however, correlation between triplets of signals is ignored; in many circuits multiple (> 2) signals reconverging at gates close to the output are strongly correlated.

In this section, we describe an approach to estimate the average power dissipation in sequential logic circuits under user-specified input sequences or programs [6]. Both temporal and spatially correlated sequences can be modeled using a finite state machine, termed an Input-Modeling Finite State Machine (IMFSM). Power estimation can be carried out using the sequential circuit power estimation methods of Section 4.3 on a cascade circuit consisting of the IMFSM and the original sequential circuit.

This technique is applicable to estimating the switching activity, and therefore power dissipation, of processors running application programs. We do not, however, model the power dissipated in external memory (e.g., DRAM, SRAM), or caches. This approach is useful in the architectural and logical design of programmable controllers and processors, because it enables the accurate evaluation of power dissipated in a controller or processor, when specific application programs are run.

Recent work in power analysis of embedded software [13] uses a different approach to estimate the power dissipated by a processor when a given program is run on the processor. An instruction-level energy model has been developed, and validated on the 486DX2. The advantages of this approach are that it is efficient and quite accurate and can take into account the power dissipated in the entire system, i.e., processor + memory + interconnect. A disadvantage is that each different architecture or different instruction set requires a significant amount of empirical analysis on implemented hardware to determine the base cost of individual instructions.

4.5.1 Completely and Incompletely Specified Input Sequences

Assume that we are given a sequential circuit M. We first consider the problem of estimating the average power dissipation in M upon the application of a periodic completely-specified input sequence C. An easy way of doing this is to perform timing simulation on the circuit for the particular vectors, and measure the activities at each gate. This, however, will become very time-consuming for incompletely-specified vector sequences.

Given the input sequence $C = \{c_1, c_2, ..., c_N\}$, we specify the State Transition Graph (STG) of an autonomous Input-Modeling Finite State Machine (IMFSM), call it A, as follows. A has N states, s_1 through s_N. For

4. POWER ESTIMATION FOR SEQUENTIAL CIRCUITS

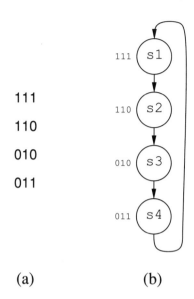

(a) (b)

Figure 4.12: Example of autonomous IMFSM for a four-vector sequence.

$1 \leq i < N$ we have a transition from s_i to s_{i+1}. Additionally we have a transition from s_N to s_1. A is a Moore machine, and the output associated with each state s_i is the corresponding completely-specified vector c_i. An example of a four-vector sequence with each vector completely-specified over three bits is given in Figure 4.12(a), and the STG of the derived IMFSM is shown in Figure 4.12(b).

A logic-level implementation of A can be obtained by arbitrarily assigning distinct codes to the states s_i, $1 \leq i \leq N$, using $\lceil \log_2 N \rceil$ bits[2]. The encoding does not affect the power estimation step as we will ignore any switching activity or power dissipation in A.

In order to estimate the average power dissipated in M upon the application of a given completely-specified input sequence C, the power estimation methods of Section 4.3 are applied to the cascade $A \rightarrow M$ depicted in Figure 4.13. Since the cascade $A \rightarrow M$ does not have any external inputs, no assumptions regarding input probabilities need to be made.

Let us now consider the problem of estimating the average power dissipation in M upon the application of a periodic incompletely-specified input sequence I. By incompletely-specified we mean that the unspecified inputs

---

[2] $\lceil x \rceil$ is the smallest integer greater or equal to x.

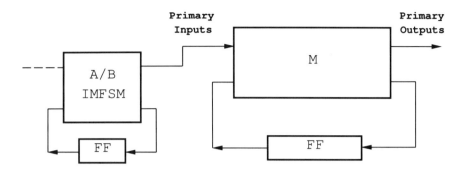

Figure 4.13: Cascade of IMFSM and given sequential circuit.

can take on either the 0 or 1 value with known probability.

As an example, consider the incompletely-specified sequence

```
11-
-1-
-01
-11
```

Completely-specified sequences contained in this sequence and that can possibly be applied to M are

```
110   111   110   110
010   111   110   111
001   101   101   001
011   111   111   011
```

among many others.

We are given the input sequence $D = \{d_1, d_2, ..., d_N\}$, over inputs $I_1, I_2, ..., I_M$. We will assume that the − entries for any I_j are uncorrelated. The − entries for each I_j have a user-specified probability of being a 1 denoted by $prob(I_j)$.

We specify the STG of the IMFSM, call it B, as follows. B has N states, s_1 through s_N, M primary inputs $I_1, I_2, ..., I_M$, and M primary outputs $o_1, o_2, ..., o_M$. For $1 \leq i < N$ we have a transition from s_i to s_{i+1} regardless of the values of the I_j's. We also have a transition from s_N to s_1 regardless of the values of the I_j's. However, B is a Mealy machine, and the output associated with each transition $s_i \rightarrow s_{i+1}$ is a logical function dependent on the corresponding d_i. An example of the incompletely-specified four-vector

4. POWER ESTIMATION FOR SEQUENTIAL CIRCUITS

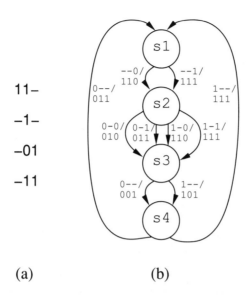

(a)   (b)

Figure 4.14: Example of Mealy IMFSM for a four-vector sequence.

sequence used above is reproduced in Figure 4.14(a), and the STG of the derived IMFSM is shown in Figure 4.14(b). Since $d_1 = 11-$, we have $o_1 = 1$, $o_2 = 1$ and $o_3 = I_3$ for the transition from s_1. Similarly for the other transitions.

As before, a logic-level implementation of B can be obtained by arbitrarily assigning distinct codes to the states s_i, $1 \leq i \leq N$, using $\lceil \log_2 N \rceil$ bits. The encoding does not affect the power estimation step.

In order to estimate the average power dissipated in M upon the application of a given incompletely-specified input sequence C, the strategies of Section 4.3 are applied to the cascade $B \rightarrow M$. The given static or transition probabilities $prob(I_j)$ of the primary inputs $I_1, I_2, ..., I_M$ to B are used to estimate the power. Note that the probabilities for all inputs to M are automatically derived.

4.5.2 Assembly Programs

In many applications, a processor receives a set of instructions as an input. An important problem is to estimate the power dissipated in the processor when it runs a given application program or a set of application programs. In this section, we describe ways of modeling an input assembly

Format	$\langle 31:26 \rangle$	$\langle 25:21 \rangle$	$\langle 20:16 \rangle$	$\langle 15:13 \rangle$	$\langle 12 \rangle$	$\langle 11:5 \rangle$	$\langle 4:0 \rangle$
Operate	Opcode	R_a	R_b	000	0	Function	R_c
Operate with literal	Opcode	R_a	Literal		1	Function	R_c
Memory	Opcode	R_a	R_b	*disp.m*			
Branch	Opcode	R_a	*disp.b*				

Instruction	Opcode	Function	Operation
add	0x10	0x20	$R_c \leftarrow \langle R_a \rangle + \langle R_b \rangle \vert Lit$
and	0x11	0x00	$R_c \leftarrow \langle R_a \rangle \wedge \langle R_b \rangle \vert Lit$
or	0x11	0x20	$R_c \leftarrow \langle R_a \rangle \vee \langle R_b \rangle \vert Lit$
sll	0x12	0x39	$R_c \leftarrow \langle R_a \rangle\ SLL\ \langle R_b \rangle \vert Lit_{5:0}$
srl	0x12	0x34	$R_c \leftarrow \langle R_a \rangle\ SRL\ \langle R_b \rangle \vert Lit_{5:0}$
sub	0x10	0x29	$R_c \leftarrow \langle R_a \rangle - \langle R_b \rangle \vert Lit$
xor	0x11	0x40	$R_c \leftarrow \langle R_a \rangle \oplus \langle R_b \rangle \vert Lit$
cmpeq	0x10	0x2D	if $\langle R_a \rangle = \langle R_b \rangle, R_c \leftarrow 1$, else $R_c \leftarrow 0$
cmple	0x10	0x6D	if $\langle R_a \rangle \leq \langle R_b \rangle, R_c \leftarrow 1$, else $R_c \leftarrow 0$
cmplt	0x10	0x4D	if $\langle R_a \rangle < \langle R_b \rangle, R_c \leftarrow 1$, else $R_c \leftarrow 0$
ld	0x29		$EA \leftarrow \langle R_b \rangle + SEXT(disp.m),$ $R_a \leftarrow MEMORY[EA]$
st	0x2D		$EA \leftarrow \langle R_b \rangle + SEXT(disp.m),$ $MEMORY[EA] \leftarrow \langle R_a \rangle$
br	0x30		$R_a \leftarrow PC, PC \leftarrow \langle PC \rangle + 4 \cdot SEXT(disp.b)$
bf	0x39		Update $PC, EA \leftarrow \langle PC \rangle + 4 \cdot SEXT(disp.b),$ if $\langle R_a \rangle = 0, PC \leftarrow EA$
bt	0x3D		Update $PC, EA \leftarrow \langle PC \rangle + 4 \cdot SEXT(disp.b),$ if $\langle R_a \rangle \neq 0, PC \leftarrow EA$

Table 4.9: α_0 **instruction set.**

program as a IMFSM so conventional sequential estimation methods can be used.

For this purpose we will focus on a simple instruction set for a RISC processor α_0, which is a subset of the instruction set for the DEC Alpha™ microprocessor. Table 4.9 gives a description of the α_0 instruction set.

Given an arbitrary α_0 program, we derive a logic-level IMFSM *B* which is cascaded with the processor as illustrated in Figure 4.13 to estimate average power consumption when the program runs on the processor. The model for the processor is illustrated in Figure 4.15. The processor is a sequential circuit consisting of a register file, arithmetic units, and control logic. It receives as input an instruction stream and reads and writes an external

4. POWER ESTIMATION FOR SEQUENTIAL CIRCUITS

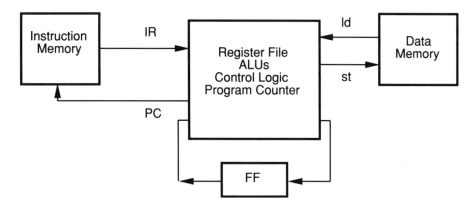

Figure 4.15: Processor model.

memory.

A key assumption that is made is that data values loaded *from* memory are random and uncorrelated. Therefore, the effect of a sequence of stores to, and loads from the same location in memory is not modeled. If we did not make this assumption then we would have to deal with the entire state space of the memory – a very difficult task. Note that in this approach we are also not concerned with the power dissipated in the external memory.

We will now describe how to generate a IMFSM given an arbitrary program comprised of a sequence of assembly instructions. Let the program P be a sequence of instructions $P = \{r_1, r_2, ..., r_N\}$. The STG of the Moore IMFSM Q has N states. For each of the different classes of instructions in Table 4.9 we show how to derive the STG of Q.

- Operate: If r_i is an Operate instruction (e.g., add, cmplt) we assign r_i as the output of state s_i. s_i makes an unconditional transition to s_{i+1}.

- Branch: If r_i is a Branch instruction, we determine the branch target instruction, call it r_j. State s_i makes a transition to state s_j if variable $v_i = 1$, and a transition to state s_{i+1} if $v_i = 0$. The probability of v_i being a 1 will be determined by preprocessing the program P as described later in the section. The output associated with s_i is r_i.

- Memory: If r_i is a Memory instruction, the output associated with s_i is r_i. On a load instruction (ld), R_a is loaded with a random value from

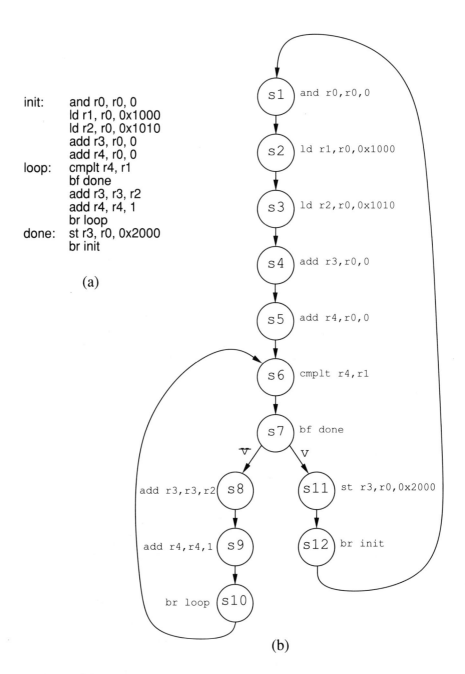

Figure 4.16: Example of Mealy IMFSM for an assembly program.

4. POWER ESTIMATION FOR SEQUENTIAL CIRCUITS

memory. The inputs to the processor from memory will have certain probabilities associated with 0 or 1 values. Since we are treating the data memory as an external memory, a store instruction (st) is essentially a null operation.

We now elaborate on the probabilities of the branch variables (v_i's). Branch prediction is a problem that has received some attention [11, pp. 103-109]. The probabilities of the branch variable $v_i = 1$ corresponds to the probability that a branch is taken on the execution of instruction r_i, and this probability can be determined, at least approximately, by preprocessing the program P.

For example, if we have a constant iteration loop with N iterations, the probability of staying in the loop is computed as $\frac{N}{N+1}$ and the probability of exiting the loop as $\frac{1}{N+1}$. If comparisons between data operands are used to determine branch conditions, the probability of the comparison evaluating to a 1 assuming random data operands can be calculated. For example, the probability that $a \geq b$ is 0.5, and the probability that $a + b > c$ is 0.75.

Additionally, we can run the program P with several different inputs, and obtain the information regarding the relative frequency with which each conditional branch is being taken versus not being taken. This relative frequency is easily converted into the probabilities for the v_i's.

As before once the STG of the IMFSM has been derived and encoded, estimation can be carried out using the topology of Figure 4.13. An example assembly program for the processor α_0 is given in Figure 4.16(a) and the STG of its corresponding IMFSM is shown in Figure 4.16(b). The average power dissipation of the processor when executing the program is computed using the estimation method of Section 4.3.

4.5.3 Experimental Results

In this section we present some experimental results obtained using the methods of Sections 4.5.1 and 4.5.2.

We compute the power dissipation of the cascade circuit consisting of the IMFSM driving the sequential circuit or processor (cf. Figure 4.13) using the techniques of Section 4.3. However, the method described in this section is not tied to a particular sequential power estimation strategy. Any strategy used has to be able to:

Circuit Name	Gate	FF	Uniform-Prob Power	Uniform-Prob CPU
bbtas	26	3	134	0.4
cse	136	4	454	13.5
keyb	174	5	587	17.5
kirkman	171	4	734	6.7
planet	333	6	2359	33.7
styr	318	5	1195	31.5
tbk	483	5	1835	81.7
train4	15	2	85	0.3
s298	119	14	441	2.5
s444	181	21	411	6.7
s526	193	21	529	5.3
s713	393	19	1176	333.7
s1196	529	18	2674	174.2
α_0-prog1	144	75	965	4.3
α_0-prog2				

Table 4.10: Comparison of power dissipation under uniform input assumption and IMFSM computation.

1. model the correlation between applied vector pairs due to the next state logic as shown in Figure 4.5, and

2. use present state probabilities or approximate using line probabilities.

In Tables 4.10 and 4.11 we present power estimation results on sequential circuits of three different types, small machines synthesized from State Transition Graph descriptions, larger controller circuits, and a small processor similar to the α_0. We give the number of gates and flip-flops in the circuit under GATES and FF respectively.

For each given sequential circuit or processor, assuming uniform primary input probabilities, we compute the power dissipation using the techniques of Section 4.3. The power estimation values assuming a clock frequency of 20MHz, a supply voltage of 5V and a unit delay model are given in the column UNIFORM-PROB of Table 4.10, together with the CPU time in seconds required for the computation on a DEC AXP 3000/500.

For the first type of circuits (for which we have a STG available) we built a transfer input sequence, i.e., an input sequence that traverses all states in

4. POWER ESTIMATION FOR SEQUENTIAL CIRCUITS

Circuit Name	IMFSM-Rand-Seq			IMFSM-Trans-Seq		
	POWER	DIFF	CPU	POWER	DIFF	CPU
bbtas	142	6.0	1.4	117	12.7	0.8
cse	473	4.2	15.5	510	12.3	15.6
keyb	479	18.4	23.4	577	1.7	21.7
kirkman	826	12.5	15.5	409	44.3	4.8
planet	2158	8.5	129.7	2147	9.0	83.5
styr	1175	1.7	46.6	1317	10.2	46.5
tbk	1705	7.1	94.1	2084	13.6	101.0
train4	54	36.5	0.4	52	38.9	0.4
s298	331	24.9	8.1	N/A		
s444	348	15.3	17.8	N/A		
s526	423	20.0	13.9	N/A		
s713	1096	6.8	513.0	N/A		
s1196	2313	13.5	197.3	N/A		
α_0-prog1	N/A			26	97.5	13.4
α_0-prog2	N/A			918	4.9	59.1

Table 4.11: Comparison of power dissipation under uniform input assumption and IMFSM computation (contd).

the STG. Additionally, for all sequential circuits we generated a random input sequence. Given these input sequences we construct an IMFSM using the methods of Section 4.5.1. The corresponding power values and CPU time are given in columns IMFSM-Trans-Seq and IMFSM-Rand-Seq of Table 4.11, respectively. Similarly, we use the techniques of Section 4.5.2 to obtain an IMFSM for two different input programs for the α_0 processor.

In Table 4.12 we give percentage errors of the present line probabilities. For each sequential circuit and each random/transfer input sequence we compute the static probabilities of the present state lines and compare them with the static probabilities obtained by assuming uniform primary input probabilities. Under MIN/MAX columns we give the percentage error of the state line with minimum/maximum static probability error. Under AVG we give the average error over all present state lines.

As we can see from Tables 4.10 and 4.11, the CPU time required to compute the power for the cascaded circuit is not much larger than that for only the original circuit. However, the power estimation error for the first set of circuits can be as high as 44%, implying that the uniform probability

Circuit Name	IMFSM-Rand-Seq			IMFSM-Trans-Seq		
	MIN	AVG	MAX	MIN	AVG	MAX
bbtas	8.1	15.7	22.4	7.3	20.3	31.7
cse	9.8	20.3	27.1	9.8	16.3	20.6
keyb	0.0	5.9	10.1	0.9	12.5	20.0
kirkman	26.9	39.6	49.3	27.1	39.7	49.3
planet	0.2	1.7	3.7	0.4	0.9	2.0
styr	8.0	20.2	29.7	13.8	19.0	22.5
tbk	1.3	3.7	5.4	0.4	12.6	17.7
train4	0.0	8.3	16.7	6.9	10.6	14.2
s298	0.0	4.9	9.5	N/A		
s444	0.0	1.6	7.4	N/A		
s526	0.0	2.9	12.8	N/A		
s713	0.0	3.3	18.3	N/A		
s1196	0.0	5.7	15.2	N/A		
α_0-prog1	N/A			0.0	2.3	49.2
α_0-prog2	N/A			0.0	0.2	1.5

Table 4.12: **Present state line probability errors.**

assumption is unrealistic. Obtaining more accurate line probabilities allows the final combinational power estimation to be more accurate. Once accurate present state line probabilities have been computed a variety of methods can be applied to estimate the power dissipated in the logic.

For the processor example, huge errors occur. The first program is a simple program which does not cause any activity in the majority of the registers and in a large fraction of the combinational logic in the processor. The difference between the average power dissipated when this program is run, and when random inputs are assumed is therefore very high. The second program is more complex, and it causes greater activity and greater power dissipation. Note that for the input programs to the processors we have assumed a random distribution for data values.

4.6 Summary

Average power dissipation estimation for sequential circuits is a difficult problem both from a standpoint of computational complexity, and from a standpoint of modeling the correlation due to feedback and correlation in input sequences.

We presented a framework for sequential power estimation in this section. In this framework, state probabilities can be computed using the Chapman-Kolmogorov equations (Section 4.2), and present state line probabilities can be computed by solving a system of non-linear equations (Section 4.3). The results presented in Section 4.4 show that the latter is significantly more efficient for medium to large circuits, and does not sacrifice accuracy. For acyclic circuits, the computation of switching activity can be done exactly and more efficiently without calculating state or state line probabilities (Section 4.1).

This framework for sequential power estimation has been implemented within SIS [12], the synthesis environment from the CAD group at the University of California at Berkeley, and is now part of their standard distribution.

This framework of power estimation for sequential circuits can be used with any power estimation technique for combinational circuits that can handle transition probabilities at the inputs. In the case of pipelines, the exact transition probabilities for the inputs to the combinational logic block of each pipeline stage are generated as shown in Figure 4.17(a). For cyclic circuits, we first use the method of Section 4.3 to compute the present state line probabilities, and then use the circuit of Figure 4.17(b) to generate the transition probabilities for the state lines.

We showed how user-specified sequences and programs can be modeled using a finite state machine, termed an Input-Modeling Finite State Machines or IMFSM (Section 4.5). Power estimation can be carried out using existing sequential circuit power estimation methods on a cascade circuit consisting of the IMFSM and the original sequential circuit.

Given input sequences or programs, we need to keep the IMFSM description reasonably compact, in order to manage the computational complexity of estimation. This implies that we need to make certain assumptions, the primary one being that data values are assumed to be uncorrelated. This assumption can be relaxed by using empirical data for particular applications such as voice and video, and we are currently looking at methods to derive this information automatically.

References

[1] R. Bahar, E. Frohm, C. Gaona, G. Hachtel, E. Macii, A. Pardo, and

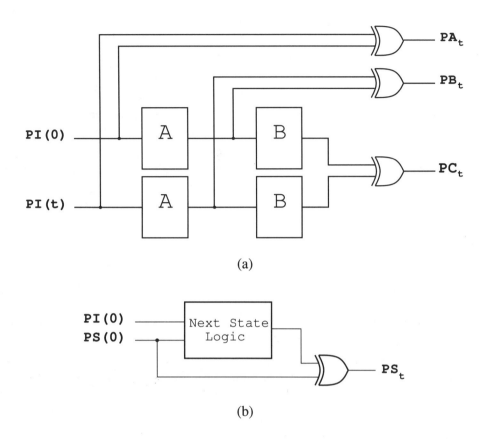

Figure 4.17: Generation of transition probabilities: (a) pipeline; (b) cyclic circuit.

F. Somenzi. Algebraic Decision Diagrams and their Applications. In *Proceedings of the International Conference on Computer-Aided Design*, pages 188–191, November 1993.

[2] A. Ghosh, S. Devadas, K. Keutzer, and J. White. Estimation of Average Switching Activity in Combinational and Sequential Circuits. In *Proceedings of the 29th Design Automation Conference*, pages 253–259, June 1992.

[3] G. Hachtel, E. Macii, A. Pardo, and F. Somenzi. Probabilistic Analysis of Large Finite State Machines. In *Proceedings of the 31st Design Automation Conference*, pages 270–275, June 1994.

[4] Z. Kohavi. *Switching and Finite Automata Theory*. Computer Science Press, 1978.

[5] R. Marculescu, D. Marculescu, and M. Pedram. Efficient Power Estimation for Highly Correlated Input Streams. In *Proceedings of the 32nd Design Automation Conference*, pages 628–634, June 1995.

[6] J. Monteiro and S. Devadas. Techniques for the Power Estimation of Sequential Logic Circuits Under User-Specified Input Sequences and Programs. In *Proceedings of the International Symposium on Low Power Design*, pages 33–38, April 1995.

[7] J. Monteiro, S. Devadas, and B. Lin. A Methodology for Efficient Estimation of Switching Activity in Sequential Logic Circuits. In *Proceedings of the 31st Design Automation Conference*, pages 12–17, June 1994.

[8] F. Najm, S. Goel, and I. Hajj. Power Estimation in Sequential Circuits. In *Proceedings of the 32nd Design Automation Conference*, pages 635–640, June 1995.

[9] J. Ortega and W. Rheinboldt. *Iterative Solution of Nonlinear Equations in Several Variables*. Academic Press, Inc., Boston, MA, 1970.

[10] A. Papoulis. *Probability, Random Variables and Stochastic Processes*. McGraw-Hill, 3rd edition, 1991.

[11] D. Patterson and J. Hennessy. *Computer Architecture: a Quantitative Approach*. Morgan Kaufman Publishers, 1990.

[12] E. Sentovich, K. Singh, C. Moon, H. Savoj, R. Brayton, and A. Sangiovanni-Vincentelli. Sequential Circuit Design Using Synthesis and Optimization. In *Proceedings of the International Conference on Computer Design: VLSI in Computers and Processors*, pages 328–333, October 1992.

[13] V. Tiwari, S. Malik, and A. Wolfe. Power Analysis of Embedded Software: A First Step Toward Software Power Minimization. *IEEE Transactions on VLSI Systems*, 2(4):437–445, December 1994.

[14] C-Y. Tsui, J. Monteiro, M. Pedram, S. Devadas, A. Despain, and B. Lin. Power Estimation for Sequential Logic Circuits. *IEEE Transactions on VLSI Systems*, 3(3):404–416, September 1995.

[15] C-Y. Tsui, M. Pedram, and A. Despain. Exact and Approximate Methods for Switching Activity Estimation in Sequential Logic Circuits. In *Proceedings of the 31st Design Automation Conference*, pages 18–23, June 1994.

Chapter 5

Optimization Techniques for Low Power Circuits

Now that we have developed tools which can efficiently estimate the average power dissipation of combinational and sequential logic circuits, we have a means of comparing different implementations of the same system, and therefore a way to direct logic synthesis tools for low power optimization.

In this chapter we present a review of previously proposed techniques for the optimization for low power of circuits described at the logic level. Recall that at this abstraction level, the model for average power dissipation is given by Equation 2.2, which we reproduce here:

$$P_i = \frac{1}{2} \cdot C_i \cdot V_{DD}^2 \cdot f \cdot N_i \qquad (5.1)$$

All optimization techniques described in this chapter assume that the clock frequency f and power supply voltage V_{DD} have been defined previously. Reducing the clock frequency is an obvious way to reduce power dissipation. However, many designers are not willing to accept the associated performance penalty. In fact, the figure of merit used by many designers is Mops/mW, million of operations per mili-Watt, and this value stays constant for different values of f.

Even better is to reduce the supply voltage, given the quadratic relationship with power. However, reducing the supply voltage increases significantly the signal propagation delays, decreasing the maximum operating frequency and thus again reducing the system's performance. In [10] the authors show that the power-delay product still reduces if we lower the supply voltage. The loss in performance can be recovered with the use of parallel processing, i.e., hardware duplication, which in turn translates higher capacitances C. Taking all these factors into account, it is possible to reach an optimum voltage level for a particular design style [10].

Given optimal f and V_{DD}, the problem of optimizing a circuit for low power is to minimize

$$\sum_i C_i \cdot N_i \qquad (5.2)$$

over all the gates in the logic circuit. This expression is often called the *switched capacitance*. Therefore we can attempt reducing the global switching activity of the circuit, reducing the global circuit capacitance or redistributing the switching in the circuit such that the switching activity of signals driving large capacitances is reduced, perhaps at the expense of increasing the switching activity of some signal driving a smaller capacitance.

In Section 5.1 we describe an important optimization method for low power: transistor sizing. While strictly this is not a gate level optimization technique, its importance has led to the incorporation of transistor sizing into logic synthesis systems. Section 5.2 is dedicated to techniques that work on restructuring the combinational logic circuit and in Section 5.3 we focus on techniques that make use of properties particular to sequential circuits.

5.1 Power Optimization by Transistor Sizing

Power dissipation is directly related to the capacitance being switched (cf. Equation 5.1). Low power designs should therefore use minimum sized transistors. However, there is a performance penalty in using minimum sized devices. The problem of *transistor sizing* is computing the sizes of the transistors in the circuit that minimizes power dissipation while still meeting the delay constraints specified for the design.

Transistor sizing for minimum area is a well established problem [26]. There is a subtle difference between this problem and sizing for low power. If the critical delay of the circuit exceeds the design specifications and thus some transistors need to be resized, methods for minimum area will focus on minimizing the total enlargement of the transistors. On the other hand, methods for low power will first resize those transistors driven by signals with lower switching activity.

5. OPTIMIZATION TECHNIQUES FOR LOW POWER CIRCUITS

A technique for transistor resizing targeting minimum power is described in [31]. Initially minimum sized devices are used. Each path whose delay exceeds the maximum allowed is examined separately. Transistors in the logic gates of these paths are resized such that the delay constraint is met. Signal transition probabilities are used to measure the power penalty of each resizing. The option with least power penalty is selected. A similar method is presented in [3]. This method is able to take false paths into account when computing the critical path of the circuit.

In [7] the authors note that the short-circuit currents are proportional to the transistor sizing. Thus the cost function used in [7] also minimizes short-circuit power.

These methods work on local optimizations. A global solution for the transistor sizing problem for low power is proposed in [6]. The problem is modeled as:

$$\tau_g = \tau_{intr} + k \frac{C_{wire} + \sum_{i \in \text{fanout}(g)} S_i C_{in,i}}{S_g} \quad (5.3)$$

$$T_g = \tau_g + \max_{i \in \text{fanin}(g)} T_i \quad (5.4)$$

$$P_g = N_g (C_{wire} + \sum_{i \in \text{fanout}(g)} S_i C_{in,i}) \quad (5.5)$$

where S_g, N_g, P_g and τ_g are respectively the sizing factor, switching activity, power dissipation and delay of gate g. τ_{intr} and k are constants representing respectively the intrinsic delay of the gate and ratio between delay and the capacitive load the gate is driving. T_g is the worst case propagation delay from an input to the output of g. C denotes load capacitances.

The solution to the optimization problem is achieved using Linear Programming (LP) [27]. A piecewise linear approximation is obtained for Equation 5.3. The constraints for the LP problem are:

$$\tau_g \geq k_{1,1} - k_{1,2} S_g + k_{1,3} \sum_i S_i C_{in,i}$$
$$\vdots \qquad \text{(from Equation 5.3)}$$
$$\tau_g \geq k_{n,1} - k_{n,2} S_g + k_{n,3} \sum_i S_i C_{in,i}$$
$$S_{min} \leq S_g \leq S_{max}$$

$$T_g \geq T_i + \tau_g, \quad \forall_{i \in \text{fanin}(g)} \qquad \text{(from Equation 5.4)}$$
$$T_{max} \geq T_g$$

and the objective function to minimize is:

$$P = \sum_{\text{over all gates } g} P_g$$

where $k_{i,j}$ are constants computed such that we get a best fit for the linearized model.

As devices shrink in size, the delay and power associated with interconnect grow in relative importance. In [12] the authors propose that wiresizing should be considered together with transistor sizing. Wider lines present less resistance but have higher capacitance. A better global solution in terms of power can be achieved if both transistor and wire sizes are considered simultaneously.

5.2 Combinational Logic Level Optimization

In this section we review techniques that work on restructuring combinational logic circuits to obtain a less power consuming circuit. The techniques we present in this section focus on reducing the switched capacitance within traditional design styles.

A different design style targeting specifically low power dissipation is proposed in [20]. It is based on *Shannon circuits* where for each computation a single input-output path is active, thus minimizing switching activity. Techniques are presented on how to keep the circuit from getting too large, as this would increase the total switched capacitance.

5.2.1 Path Balancing

Spurious transitions account for a significant fraction of the switching activity power in typical combinational logic circuits [28, 14]. In order to reduce spurious switching activity, the delay of paths that converge at each gate in the circuit should be roughly equal. Solutions to this problem, known as *path balancing*, have been proposed in the context of *wave-pipelining* [19]. One technique involves restructuring the logic circuit, as illustrated in Figure 5.1. Additionally, by selectively inserting unit-delay buffers to the inputs of gates in a circuit, the delays of all paths in the circuit can be made equal (Figure 5.2). This addition will not increase the critical delay of the circuit, and will effectively eliminate spurious transitions. However, the addition of buffers increases capacitance which may offset the reduction in switching activity.

5. OPTIMIZATION TECHNIQUES FOR LOW POWER CIRCUITS

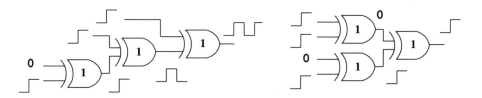

Figure 5.1: Logic restructuring to minimize spurious transitions.

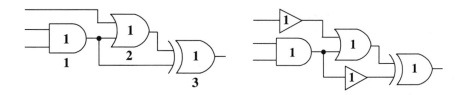

Figure 5.2: Buffer insertion for path balancing.

5.2.2 Don't-care Optimization

Multilevel circuits are optimized by repeated two-level minimization with appropriate don't-care sets. Consider the circuit of Figure 5.3. The structure of the logic circuit may imply some combinations over nodes A, B and C never occur. These combinations form the *Controllability* or *Satisfiability Don't-Care Set* (SDC) of F. Similarly, there may be some input combinations for which the value of F is not used in the computation of the outputs of the circuit. The set of these combinations is called the *Observability Don't-Care Set* (ODC) [13, pp. 178-179].

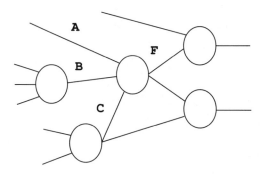

Figure 5.3: SDCs and ODCs in a multilevel circuit.

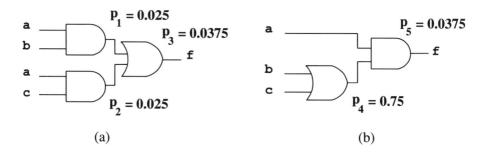

Figure 5.4: Logic factorization for low power.

Traditionally don't-care sets have been used for area minimization [4]. Recently techniques have been proposed (e.g., [28, 16]) for the use of don't-cares to reduce the switching activity at the output of a logic gate. The transition probability of the output f of a static CMOS gate is given by $2prob(f)(1 - prob(f))$ (ignoring temporal correlation). The maximum for this function occurs when $prob(f) = 0.5$. The authors of [28] suggest including minterms in the don't-care set in the ON-set of the function if $prob(f) > 0.5$ or in the OFF-set if $prob(f) < 0.5$. In [16] this method is extended to take into account the effect that the optimization of a gate has in the switching probability of its transitive fanout.

5.2.3 Logic Factorization

A primary means of technology-independent optimization (i.e., before technology mapping) is the factoring of logical expressions. For example, the expression $(a \land c) \lor (a \land d) \lor (b \land c) \lor (b \land d)$ can be factored into $(a \lor b) \land (c \lor d)$, reducing transistor count considerably. Common subexpressions can be found across multiple functions and reused. Kernel extraction is a commonly used algorithm to perform multilevel logic optimization for area [8]. In this algorithm, the kernels of given expressions are generated and those kernels that maximally reduce the literal count are selected.

When targeting power dissipation, the cost function is not literal count but switching activity. Even though transistor count may be reduced by factorization, the total switched capacitance may increase. Consider the example shown in Figure 5.4 and assume that a has a low probability $prob(a) = 0.1$ and b and c have each $prob(b) = prob(c) = 0.5$. The total switched capacitance in the circuit of Figure 5.4(a) is $2(2prob(a)(1 - prob(a)) + prob(b)(1 - prob(b)) +$

5. OPTIMIZATION TECHNIQUES FOR LOW POWER CIRCUITS

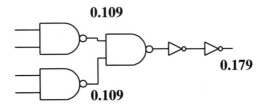

Figure 5.5: Circuit to be mapped, with switching activity information.

GATE	AREA	INTRINSIC CAPACITANCE	INPUT LOAD CAPACITANCE
INV	928	0.1029	0.0514
NAND2	1392	0.1421	0.0747
AOI22	2320	0.3410	0.1033

Figure 5.6: Information about the technology library.

$prob(c)(1 - prob(c)) + p_1(1 - p_1) + p_2(1 - p_2) + p_3(1 - p_3))C = 1.52C$ and in the circuit of Figure 5.4(b) is $2(prob(a)(1 - prob(a)) + prob(b)(1 - prob(b)) + prob(c)(1 - prob(c)) + p_4(1 - p_4) + p_5(1 - p_5))C = 1.61C$. Clearly factorization is not always desirable in terms of power. Further, kernels that lead to minimum literal count do not necessarily minimize the switched capacitance.

Modified kernel extraction methods that target power are described in [25, 22, 17, 24]. The algorithms proposed compute the switching activity associated with the selection of each kernel. Kernel selection is based on the reduction of both area and switching activity.

5.2.4 Technology Mapping

Technology mapping is the process by which a logic circuit is implemented in terms of the logic elements available in a particular technology library. Associated with each logic element is an area and a delay cost. The traditional optimization problem is to find the implementation that meets some delay constraint and minimizes the total area cost. Techniques to efficiently find an optimal solution to this problem have been proposed [18].

As long as the delay constraints are still met, the designer is usually willing to make some tradeoff between area and power dissipation. Consider the circuit of Figure 5.5. Mapping this circuit for minimum area using the technology library presented in Figure 5.6 yields the circuit presented in

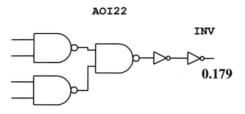

Area = 2320 + 928 = 3248
Power = 0.179 × (0.3410 + 0.0514) + 0.179 × (0.1029) = 0.0887

Figure 5.7: Mapping for minimum area.

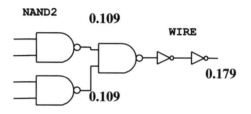

Area = 1392 × 3 = 4176
Power = 0.109 × (0.1421 + 0.0747) × 2 + 0.179 × (0.1421) = 0.0726

Figure 5.8: Mapping for minimum power.

Figure 5.7. The designer may prefer to give up some area in order to obtain the more power efficient design of Figure 5.8.

The graph covering formulation of [18] has been extended to use switched capacitance as part of the cost function. The main strategy to minimize power dissipation is to hide nodes with high switching activity within complex logic elements as capacitances internal to gates are generally much smaller. Although using different models for delay and switching activity estimation, techniques such as those described in [32, 35, 21] all use this approach to minimize power dissipation during technology mapping.

Most technology libraries include the same logic element with different sizes (i.e., drive capability). Thus, in technology mapping for low power, the choice of the size of each logic element such that the delay constraints are met with minimum power consumption is made. This problem is the discrete counterpart of the transistor sizing problem of Section 5.1 and is addressed in [31, 2, 30].

5.3 Sequential Optimization

We now focus on techniques for low power that are specific to synchronous sequential logic circuits. A characteristic of this type of circuits is that switching activity is easily controllable by deciding whether or not to load new values to registers. Further, at the output of registers we always have a clean transition, free from glitches.

5.3.1 State Encoding

State encoding is the process by which a unique binary code is assigned to each state in a Finite State Machine (FSM). Although this assignment does not influence the functionality of the FSM, it determines the complexity of the combinational logic block in the FSM implementation (cf. Figure 4.3).

State encoding for minimum area is a well-researched problem [1, Chapter 5]. The optimum solution to this problem has been proven to be NP-hard. Heuristics that work well assign codes with minimum Hamming distances to states that have edges connecting them in the State Transition Graph (STG). This potentially enables the existence of larger kernels or kernels that can be used a larger number of times.

Targeting low power dissipation, the heuristics go one step further: assign minimum Hamming distance codes to states that are connected by edges that have higher probability of being traversed. The probability that a given edge in the STG is traversed is given by the steady-state probability of the STG being in the start state of the edge times the static probability of the input combination associated with that edge (cf. Equation 4.1). Whenever this edge is exercised, only a small number of state lines (ideally one) will change, leading to reduced overall switching activity in the combinational logic block. This is the cost function used in the techniques proposed in [25, 23, 15].

In [34], the technique takes into account not only the power in the state lines but also in the combinational logic by using in the cost function the savings relative to cubes possible to obtain for a given state encoding.

5.3.2 Encoding in the Datapath

Encoding to reduce switching activity in datapath logic has also been the subject of attention. A method to minimize the switching on buses is proposed in [29]. Buses usually correspond to long interconnect lines and therefore have a very high capacitance. Thus any reduction in the switching activity of a bus may correspond to significant power savings. In [29], an extra line E is added to the bus which indicates if the value being transferred is the true value or needs to be bitwise complemented upon receipt. Depending on the value transferred in the previous cycle, a decision is made to either transfer the true current value or the complemented current value, so as to minimize the number of transitions in the bus lines. For example, if the previous value transferred was 0000, and the current value is 1011, then the value 0100 is transferred, and the line E is asserted to signify that the value 0100 has to be complemented at the other end. The number of lines switching in the bus has been reduced from three to two. Other methods of bus coding are also proposed in [29].

Methods to implement arithmetic units other than in standard two's complement arithmetic are also being investigated. A method of one-hot residue coding to minimize switching activity of arithmetic logic is presented in [11].

5.3.3 Gated Clocks

Large VLSI circuits such as processors contain register files, arithmetic units and control logic. The register file is typically not accessed in each clock cycle. Similarly, in an arbitrary sequential circuit, the values of particular registers need not be updated in every clock cycle. If simple conditions that determine the inaction of particular registers can be computed, then power reduction can be obtained by gating the clocks of these registers [9] as illustrated in Figure 5.9. When these conditions are satisfied, the switching activity within the registers is reduced to negligible levels. Detection and shut-down of unused hardware is done automatically in current generations of Pentium and PowerPC processors. The Fujitsu SPARClite™ processor provides software controls for shutting down hardware.

The same method can be applied to "turn off" or "power down" arithmetic units when these units are not in use in a particular clock cycle. For example, when a branch instruction is being executed by a CPU, a multiply

5. OPTIMIZATION TECHNIQUES FOR LOW POWER CIRCUITS

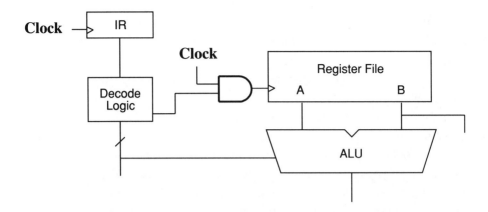

Figure 5.9: Reducing switching activity in the register file and ALU by gating the clock.

unit may not be used. The input registers to the multiplier are maintained at their previous values, ensuring that switching activity power in the multiplier is zero for this clock cycle.

In [5] a gated clock scheme applicable to FSMs is proposed. The clock to the FSM is turned off when the FSM is in a state with a self loop waiting for some external condition to arrive. Techniques to transform locally a Mealy machine into a Moore machine are presented so that the opportunity for gating the clock is increased.

As a follow up of the *precomputation* method that is presented in Chapter 7, a technique called *guarded evaluation* [33] achieves data-dependent power down at the sequential logic level. Instead of adding the precomputation logic to generate the clock disabling signal, this technique uses signals already existing in the circuit to prevent transitions from propagating. Disabling signals and subcircuits to be disabled are determined by using observability don't-care sets.

5.4 Summary

We have reviewed recently proposed optimization methods for low power that work at the transistor and logic levels.

To reduce the switched capacitance, transistors should be as small as possible. For most designs we cannot afford to use only minimum sized transistors since we need to meet some performance constraints. Transistor sizing for low power enlarges first those transistors driven by signals with lower switching activity.

Combinational techniques such as don't care optimization, logic factorization and technology mapping try to reduce the switching activity in the circuit or redistribute the switching activity such that we have fewer transitions for signals driving large capacitive loads.

Other techniques that focus on reducing spurious transitions, such as path balancing are inherently limited as they do not address the zero-delay switching activity. However these techniques are independent improvements and can be used together with the other optimization techniques. The technique we describe in Chapter 6 focus on reducing spurious transitions by repositioning the registers in the circuit.

Shut-down techniques applied to sequential circuits such as those described in Section 5.3.3 have a greater potential for reducing the overall switching activity in logic circuits. We describe two optimization techniques based on the observations of Section 5.3.3, which are presented in Chapter 7 and Section 8.2.2.

References

[1] P. Ashar, S. Devadas, and A. R. Newton. *Sequential Logic Synthesis*. Kluwer Academic Publishers, Boston, Massachusetts, 1991.

[2] R. Bahar, H. Cho, G. Hachtel, E. Macii, and F. Somenzi. An Application of ADD-Based Timing Analysis to Combinational Low Power Synthesis. In *Proceedings of the International Workshop on Low Power Design*, pages 39–44, April 1994.

[3] R. Bahar, G. Hachtel, E. Macii, and F. Somenzi. A Symbolic Method to Reduce Power Consumption of Circuits Containing False Paths. In *Proceedings of the International Conference on Computer-Aided Design*, pages 368–371, November 1994.

[4] K. Bartlett, R. Brayton, G. Hachtel, R. Jacoby, C. Morrison, R. Rudell, A. Sangiovanni-Vincentelli, and A. Wang. Multi-Level Logic Minimization Using Implicit Don't Cares. *IEEE Transactions on Computer-Aided Design*, 7(6):723–740, June 1988.

[5] L. Benini and G. De Micheli. Transformation and Synthesis of FSMs for Low Power Gated Clock Implementation. In *Proceedings of the International Symposium on Low Power Design*, pages 21–26, April 1995.

[6] M. Berkelaar and J. Jess. Computing the Entire Active Area/Power Consumption versus Delay Trade-off Curve for Gate Sizing with a Piecewise Linear Simulator. In *Proceedings of the International Conference on Computer-Aided Design*, pages 474–480, November 1994.

[7] M. Borah, R. Owens, and M. Irwin. Transistor Sizing for Minimizing Power Consumption of CMOS Circuits under Delay Constraint. In *Proceedings of the International Symposium on Low Power Design*, pages 167–172, April 1995.

[8] R. Brayton, R. Rudell, A. Sangiovanni-Vincentelli, and A. Wang. MIS: A Multiple-Level Logic Optimization System. *IEEE Transactions on Computer-Aided Design*, 6(6):1062–1081, November 1987.

[9] A. Chandrakasan. *Low-Power Digital CMOS Design*. PhD thesis, University of California at Berkeley, UCB/ERL Memorandum No. M94/65, August 1994.

[10] A. Chandrakasan, M. Potkonjak, R. Mehra, J. Rabaey, and R. Broderson. Optimizing Power Using Transformations. *IEEE Transactions on Computer-Aided Design*, 14(1):12–31, January 1995.

[11] W. Chren. Low Delay-Power Product CMOS Design Using One-Hot Residue Coding. In *Proceedings of the International Symposium on Low Power Design*, pages 145–150, April 1995.

[12] J. Cong and C. Koh. Simultaneous Driver and Wire Sizing for Performance and Power Optimization. *IEEE Transactions on VLSI Systems*, 2(4):408–425, December 1994.

[13] S. Devadas, A. Ghosh, and K. Keutzer. *Logic Synthesis*. McGraw Hill, New York, NY, 1994.

[14] M. Favalli and L. Benini. Analysis of Glitch Power Dissipation in CMOS ICs. In *Proceedings of the International Symposium on Low Power Design*, pages 123–128, April 1995.

[15] G. Hachtel, M. Hermida, A. Pardo, M. Poncino, and F. Somenzi. Re-Encoding Sequential Circuits to Reduce Power Dissipation. In *Proceedings of the International Conference on Computer-Aided Design*, pages 70–73, November 1994.

[16] S. Iman and M. Pedram. Multi-Level Network Optimization for Low Power. In *Proceedings of the International Conference on Computer-Aided Design*, pages 371–377, November 1994.

[17] S. Iman and M. Pedram. Logic Extraction and Factorization for Low Power. In *Proceedings of the 32nd Design Automation Conference*, pages 248–253, June 1995.

[18] K. Keutzer. DAGON: Technology Mapping and Local Optimization. In *Proceedings of the 24th Design Automation Conference*, pages 341–347, June 1987.

[19] T. Kim, W. Burleson, and M. Ciesielski. Logic Restructuring for Wave-pipelined Circuits. In *Proceedings of the International Workshop on Logic Synthesis*, 1993.

[20] L. Lavagno, P. McGeer, A. Saldanha, and A. Sangiovanni-Vincentelli. Timed Shannon Circuits: A Power-Efficient Design Style and Synthesis Tool. In *Proceedings of the 32nd Design Automation Conference*, pages 254–260, June 1995.

[21] B. Lin. Technology Mapping for Low Power Dissipation. In *Proceedings of the International Conference on Computer Design: VLSI in Computers and Processors*, October 1993.

[22] R. Murgai, R. Brayton, and A. Sangiovanni-Vincentelli. Decomposition of Logic Functions for Minimum Transition Activity. In *Proceedings of the International Workshop on Low Power Design*, pages 33–38, April 1994.

[23] E. Olson and S. Kang. Low-Power State Assignment for Finite State Machines. In *Proceedings of the International Workshop on Low Power Design*, pages 63–68, April 1994.

[24] R. Panda and F. Najm. Technology Decomposition for Low-Power Synthesis. In *Proceedings of the Custom Integrated Circuit Conference*, 1995.

[25] K. Roy and S. Prasad. Circuit Activity Based Logic Synthesis for Low Power Reliable Operations. *IEEE Transactions on VLSI Systems*, 1(4):503–513, December 1993.

[26] S. Sapatnekar, V. Rao, P. Vaidya, and S. Kang. An Exact Solution to the Transistor Sizing Problem for CMOS Circuits Using Convex Optimization. *IEEE Transactions on Computer-Aided Design*, 12(11):1621–1634, November 1993.

[27] A. Schrijver. *Theory of Linear and Integer Programming*. Wiley, 1987.

[28] A. Shen, S. Devadas, A. Ghosh, and K. Keutzer. On Average Power Dissipation and Random Pattern Testability of Combinational Logic Circuits. In *Proceedings of the International Conference on Computer-Aided Design*, pages 402–407, November 1992.

[29] M. Stan and W. Burleson. Limited-Weight Codes for Low-Power I/O. In *Proceedings of the International Workshop on Low Power Design*, pages 209–214, April 1994.

[30] Y. Tamiya, Y. Matsunaga, and M. Fujita. LP-based Cell Selection with Constraints of Timing, Area and Power Consumption. In *Proceedings of the International Conference on Computer-Aided Design*, pages 378–381, November 1994.

[31] C. Tan and J. Allen. Minimization of Power in VLSI Circuits Using Transistor Sizing, Input Ordering, and Statistical Power Estimation. In *Proceedings of the International Workshop on Low Power Design*, pages 75–80, April 1994.

[32] V. Tiwari, P. Ashar, and S. Malik. Technology Mapping for Low Power. In *Proceedings of the 30th Design Automation Conference*, pages 74–79, June 1993.

[33] V. Tiwari, P. Ashar, and S. Malik. Guarded Evaluation: Pushing Power Management to Logic Synthesis/Design. In *Proceedings of the International Symposium on Low Power Design*, pages 221–226, April 1995.

[34] C-Y. Tsui, M. Pedram, C-A. Chen, and A. Despain. Low Power State Assignment Targeting Two- and Multi-level Logic Implementations. In

Proceedings of the International Conference on Computer-Aided Design, pages 82–87, November 1994.

[35] C-Y. Tsui, M. Pedram, and A. Despain. Technology Decomposition and Mapping Targeting Low Power Dissipation. In *Proceedings of the 30th Design Automation Conference*, pages 68–73, June 1993.

Chapter 6

Retiming for Low Power

The operation of *retiming* consists of repositioning the registers in a sequential circuit, while maintaining its external functional behavior. Retiming was first proposed in [3] as a technique to improve throughput by moving the registers in a circuit.

In this chapter, we explore the application of retiming techniques to modify the switching activity in internal signals of a circuit [6] and demonstrate the impact of these techniques on average power dissipation. The use of retiming to minimize switching activity is based on the observation that the output of registers have significantly fewer transitions than the register inputs. In particular, no glitching is present.

Consider the circuit of Figure 6.1(a). If the average switching activity at the output of gate g is N_g and the load capacitance is C_L, then the power dissipated at the output of this gate is proportional to $N_g \cdot C_L$. Now consider the situation when a flip-flop R is added to the output of g, as illustrated in Figure 6.1(b). The power dissipated by the circuit is now proportional to $N_g \cdot C_R + N_R \cdot C_L$, where N_g is as before, C_R is the capacitance seen at the input to the flip-flop, and N_R is the average switching activity at the flip-flop output. The main observation here is that $N_R < N_g$, since the flip-flop output will make at most one transition at the beginning of the clock cycle. For example, the gate g may glitch and make three transitions as shown in the figure, but the flip-flop output will make at most one transition when the clock is asserted. This implies that is possible that $N_g \cdot C_R + N_R \cdot C_L$ is less than $N_g \cdot C_L$ if both N_g and C_L are high. Thus, the addition of flip-flops to a circuit may actually decrease power dissipation. Since adding flip-flops to a circuit is a common

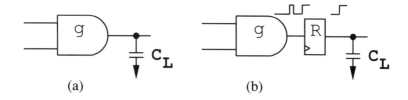

Figure 6.1: Adding a flip-flop to a circuit.

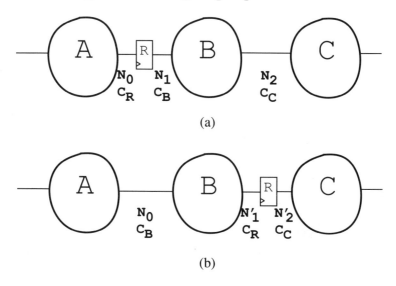

Figure 6.2: Moving a flip-flop in a circuit.

way to improve the performance of a circuit by pipelining it, it is worthwhile to exploit all the ramifications of this observation.

Next, consider the more complex scenario of altering the position of a flip-flop in a sequential circuit. Consider the circuit of Figure 6.2(a). The power dissipated by this circuit is proportional to $N_0 \cdot C_R + N_1 \cdot C_B + N_2 \cdot C_C$. Similarly, the power dissipated by the circuit of Figure 6.2(b) is proportional to $N_0 \cdot C_B + N'_1 \cdot C_R + N'_2 \cdot C_C$. One circuit may have significantly lower power dissipation than the other. Due to glitching, N'_1 may be greater than N_1 but by the same token N'_2 may be less than N_2. The capacitances of the logic blocks and the flip-flops along with the switching activities will determine which of the circuits is more desirable from a power standpoint. The circuits may also have differing performance.

We use the above observations in a heuristic retiming strategy that

6. RETIMING FOR LOW POWER

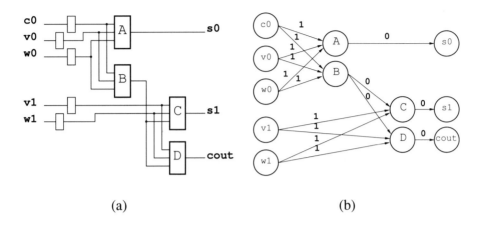

Figure 6.3: Pipelined 2-bit adder: (a) Circuit; (b) Graph.

targets power dissipation as its primary cost function. We describe this technique in Section 6.2. We begin by making a brief review of retiming in Section 6.1. Experimental results are presented in Section 6.3.

6.1 Review of Retiming

6.1.1 Basic Concepts

For the formulation of the retiming problem, a sequential circuit is generally modeled as a Directed Acyclic Graph $G(V,E,W)$, where: V is the set of vertices, with one vertex for each primary input, each primary output and each gate in the circuit; E is the set of edges, which represent the interconnections between the gates; W is a set of weights associated with each edge in the graph and it represents the number of registers in the connection corresponding to each edge. Figures 6.3 and 6.4 show two different sequential circuits and their respective graphs.

A *path* between two vertices in the graph $v_1, v_2 \in V$, $v_1 \rightsquigarrow v_2$, is defined as the sequence of edges from v_1 to v_2. The weight of the path is the sum of the weights of the edges in the path. In the particular case of k-pipelines, the weight of any path from a primary input to a primary output is always k.

The retiming operation is defined at a vertex level. The retiming of vertex v, $r(v)$, is the number of registers to be moved from the fanout edges of

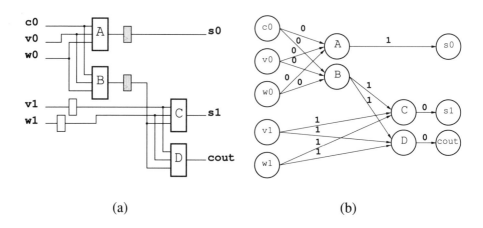

Figure 6.4: Retimed 2-bit adder: (a) Circuit; (b) Graph.

vertex v to its fanin edges. The weight $w'(e)$ after a retiming operation of an edge e from v_1 to v_2, $v_1 \xrightarrow{e} v_2$, is given by

$$w'(e) = w(e) + r(v_2) - r(v_1) \qquad (6.1)$$

It is shown in [3] that the input/output behavior of the circuit is preserved (*legal retiming*) if the retiming verifies the following conditions:

(i) $r(v) = 0$, if v is a primary input or primary output.
(ii) $w(e) + r(v_2) - r(v_1) \geq 0$, where e is an edge from vertex v_1 to vertex v_2, $v_1 \xrightarrow{e} v_2$.

Condition (i) implies that if the clock cycle in which inputs/outputs are to arrive/be available is to be maintained, then no registers should be borrowed from (or lent to) outside circuitry. Condition (ii) ensures that there are no edges in the graph with negative weights.

The circuit in Figure 6.4 is a retimed version of that in Figure 6.3. Vertices A and B were retimed by $r(A) = r(B) = -1$, the registers at the inputs of the corresponding gates in the circuit were moved to their outputs. It can be observed that the logic function performed by these circuits as seen from the outside is exactly the same. Although gates A and B are computed one clock cycle earlier in the second circuit, the outputs of the circuit are available in the same clock cycle as before.

6.1.2 Applications of Retiming

Retiming based algorithms have been used previously in logic design optimization, both targeting performance [3, 5, 1] and area [4].

In [3, 5, 1], registers are redistributed so as to minimize the delay of the longest path, thus allowing the circuit to operate at higher clock speeds.

In [4], retiming is used to allow optimization methods for combinational circuits to be applied across register boundaries. The circuit is retimed so that registers are moved to the border of the circuit, logic minimization methods are applied to the whole combinational logic block and lastly the registers are again redistributed in the circuit to maximize throughput.

In the next section we present an algorithm that applies retiming with a different cost function. We retime sequential circuits so as to minimize the power dissipated in the circuit by minimizing its switching activity [6]. This work has been extended in [2]. In [2] each register is replaced by two level-sensitive latches, each working on a different clock phase. The retiming is performed only on latches clocked on one of the clock phases. Since the latches for the other clock phase stay fixed, the state variables at the output of these latches remain the same. One of the advantages of this is that the testability characteristics of the original edge-triggered circuit and of the retimed level-sensitive circuit are the same.

6.2 Retiming for Low Power

Retiming algorithms that minimize clock periods [3, 5] rely on the fact that delay varies linearly under retiming. The delay from v_1 to v_2 is the sum of the delays in the path $v_1 \rightsquigarrow v_2$. Unfortunately that is not so for switching activity.

The retiming of a single vertex can dramatically change the switching activity in the circuit and it is very difficult to predict what this change will be. On the other hand, re-estimating the switching activity after each retiming operation is not a viable alternative as the estimation process is itself computationally very expensive.

The algorithm we propose for reducing power dissipation in a pipelined circuit heuristically selects the set of gates which, by having a flip-flop placed at their outputs, lead to the minimization of switching activity in the circuit. Gates are selected based on the amount of glitching that is present at

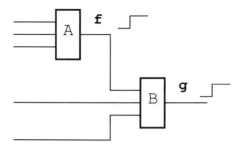

Figure 6.5: Sensitivity calculation.

their outputs and on the probability that this glitching will propagate through their transitive fanouts.

The number of registers in the final circuit can also have a high impact on the power dissipation. As a second objective, we minimize the number of registers in the circuit by performing retiming operations provided they maintain the registers previously placed and do not increase the maximum delay path.

6.2.1 Cost Function

We start by estimating the average switching activity in the combinational circuit (ignoring the flip-flops), both with zero delay (N_{zeroD}) and actual delay (N_{actD}). We compute the amount of glitching (N_{glitch}) at each gate by taking the difference of the expected number of transitions in these two cases ($N_{glitch} = N_{actD} - N_{zeroD}$).

We then evaluate the probability that a transition at each gate propagates through its transitive fanout. For each gate g in the transitive fanout of f, as in Figure 6.5, we calculate the probability of having a transition at gate g caused by a transition at gate f (*sensitivity* of gate g relative to gate f, $s_{g,f}$):

$$s_{g,f} = prob(g\updownarrow \,|f\updownarrow) = \frac{prob(f\updownarrow \wedge g\updownarrow)}{prob(f\updownarrow)} \quad (6.2)$$

where $prob(f\updownarrow)$ is the probability of a transition at the output of gate f,

$$\begin{aligned} prob(f\updownarrow) &= N_f = prob^{01}(f) + prob^{10}(f) \\ &= prob(\overline{f(0)}f(t)) + prob(f(0)\overline{f(t)}) \end{aligned} \quad (6.3)$$

6. RETIMING FOR LOW POWER

$prob(f \updownarrow \wedge g \updownarrow)$ can be obtained by computing the primary input conditions under which a transition at f triggers a transition at g:

$$\begin{aligned} prob(f \updownarrow \wedge g \updownarrow) &= prob(f(0)\overline{f(t)}\, g(0)\,\overline{g(t)}) \\ &+ prob(\overline{f(0)}f(t)\, g(0)\,\overline{g(t)}) \\ &+ prob(f(0)\overline{f(t)}\,\overline{g(0)}\, g(t)) \\ &+ prob(\overline{f(0)}f(t)\,\overline{g(0)}\, g(t)) \end{aligned} \quad (6.4)$$

We have $prob(f(0)\,g(0)) = prob(f(t)\,g(t)) = prob(f\,g)$. Therefore Equation 6.4 becomes

$$prob(f \updownarrow \wedge g \updownarrow) = 2(p_1 p_4 + p_2 p_3) \quad (6.5)$$

where

$$p_1 = prob(f\,g),\ p_2 = prob(\overline{f}\,g),\ p_3 = prob(f\,\overline{g}),\ p_4 = prob(\overline{f}\,\overline{g})$$

BDDs that represent all primary input conditions under which f and g make a transition can be constructed using the methods of Chapter 3. Computing the Boolean AND of these BDDs gives us the primary input conditions for $f \updownarrow \wedge g \updownarrow$. The probability $prob(f \updownarrow \wedge g \updownarrow)$ can be calculated using a bottom-up traversal of the BDD. Also, calculating signal transition probability at each gate ($prob(f \updownarrow)$) can be calculated using the zero delay power estimation methods in Chapter 3.

Since the objective is to reduce power, we weight these sensitivities with the capacitive load of the corresponding gate. The measure of the amount of power dissipation that is reduced by placing a flip-flop at the output of a gate f is:

$$power\_red(f) = N_{glitch}(f) \times (C_f + \sum_{g \in \text{fanout}(f)} (s_{g,f} \times C_g)) \quad (6.6)$$

The transitive fanout of a gate may contain a very large number of elements, so we restrict the number of levels in the transitive fanout that are taken into account. This not only reduces computation time, but also can increase the accuracy since glitching can be filtered out by the inertial delay of combinational logic. From empirical observations, we have concluded that computing the sensitivity of gates up to two levels down in the transitive fanout is sufficient.

One other factor that can significantly contribute to power dissipation is the number of flip-flops in the circuit. We try to minimize this number by giving higher weights to vertices with larger number of inputs ($n_i(f)$) and outputs ($n_o(f)$). A flip-flop placed at one of these vertices will be in a larger number of paths, reducing the total number of flip-flops needed. Therefore, the final cost function that we want to maximize is given by:

$$weight(f) = power\_red(f) \times (n_i(f) + n_o(f)) \qquad (6.7)$$

6.2.2 Verifying a Given Clock Period

Although we aim at the circuit that dissipates the least possible power, we might also want to set a constraint on performance by specifying the desired clock cycle of the retimed circuit.

In the retiming algorithm we will be selecting vertices that should have a flip-flop placed at the output. We restrict the selection process to vertices that still allow the retimed circuit to be clocked with the given clock period. Since the algorithm works with pipelines (acyclic circuits), this is accomplished simply by discarding vertices that have a path longer than the desired clock period, both from any primary input or to any primary output.

6.2.3 Retiming Constraints

The objective is to select the vertices (from those in the conditions of the previous section) with the highest weights, as given by Equation 6.7. The retiming constraint is that the number of selected vertices that share any input-output path should not surpass a given value (which is the number of flip-flop stages in the pipeline). The set of vertices that verify this constraint and corresponding to the highest sum of weights is chosen.

We restrict our algorithm to place one stage of flip-flops at a time. The reason for this is that, if we allowed two stages, the algorithm could select a gate f and one of its immediate fanout gates g for a set. Choosing f will eliminate most of the glitching present at g, possibly changing significantly the weight of g. This new weight of g is very difficult to predict. Thus, for pipelines with more than one stage, we apply the algorithm iteratively.

Hence the goal is to find the set of vertices with no more than one vertex per input-output path and with the highest sum of weights. The algorithm uses a binary tree search over all the vertices, keeping record of the best set so far. For large circuits, we limit the search to the most promising vertices.

6. RETIMING FOR LOW POWER

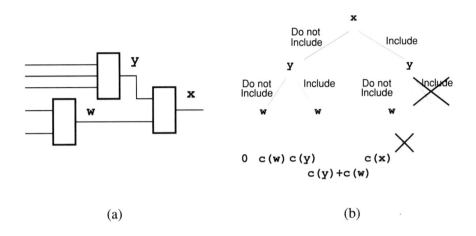

Figure 6.6: Vertex selection: (a) Circuit; (b) Binary tree.

First we check for pairwise compatibility. For each pair of vertices we check if there is one input-output path to which they both belong. This greatly simplifies the test at each level of the binary tree as we just verify if the vertex corresponding to this level is *incompatible* with any other vertex previously selected.

To exemplify this process, consider the circuit of Figure 6.6(a). We have represented in Figure 6.6(b) the binary tree for vertex selection. Right branches in the tree correspond to the vertex being selected having a flip-flop at its output and left branches to no flip-flop at the output of the vertex. Since vertex x shares input-output paths with both vertices y and w, selecting x implies that none of the other two vertices can be selected. After building the binary tree, we are left with valid combinations of vertices. The one with the highest sum of cost functions is chosen.

6.2.4 Executing the Retiming

Initially we position the flip-flops at the primary inputs of the circuit. To place a flip-flop at the output of a gate in the selected set, we recursively perform backward retiming on the vertex, adding a flip-flop at its output and removing a flip-flop from each input. This operation is repeated for vertices that have *negative* flip-flops at their output due to previous retimings. Eventually we reach the primary inputs where flip-flops are present, thereby ending the

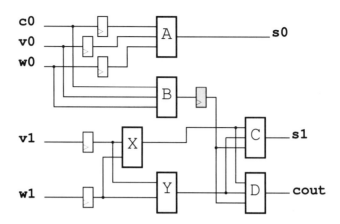

Figure 6.7: Circuit with the gates in the selected set retimed.

recursion.

Once we have placed flip-flops at the output of all the gates in the set, there are typically some flip-flops that can still be moved without disturbing the flip-flops already placed. These are flip-flops on paths that do not contain any vertex in the selected set. For instance, consider the circuit depicted in Figure 6.7 which has been through the first phase of retiming, where the only vertex in the selected set was vertex B.

The first observation is that although vertex B was retimed (and has a flip-flop at its output as was the objective), A was not. Thus the flip-flops at the inputs $c0$, $v0$ and $w0$ were not removed. In this case it is obvious that it is preferable to retime vertex A so that we reduce the number of flip-flops in the circuit (one at the output of A instead of three at the inputs).

The second observation is that the flip-flops at inputs $v1$ and $w1$ were also not touched. Vertices X and Y can be retimed and this would reduce the levels of combinational logic in the circuit from two to one. Note that retiming X and Y will make C and D retimable, but we do not allow this operation since that would remove the flip-flop from the output of B.

Thus, in the last phase of the algorithm we go through the circuit, from primary inputs to primary outputs, performing a backward retiming on retimable vertices so that:

(i) the delay does not increase over the desired clock period
(ii) the number of flip-flops is reduced

6. RETIMING FOR LOW POWER

Circuit	Retime-Delay			Retime-Power			
Name	FF	Delay	Power	FF	Delay	Power	% Red
cla_16	48	12	2389	43	12	2147	10.1
rpl_16	33	18	2303	32	32	2074	9.9
cbp_16	38	22	2748	34	42	2388	13.1
cbp_32	74	42	5590	61	71	4725	15.5
mult4	14	5	900	11	7	853	5.2
mult6	29	8	2803	22	11	2596	7.4
mult8	46	11	6104	37	15	5834	4.4

Table 6.1: Results of retiming for low power with no timing constraints.

(iii) this retiming operation does not disturb the flip-flops placed at the output of the vertices in the selected set

6.3 Experimental Results

We present results obtained by using the retiming method of Section 6.2 that directly targets power dissipation. In Table 6.1 we present the delay, in nano-seconds, and power, in micro-Watt, dissipated by circuits retimed for minimum delay, and the delay and power dissipated by circuits retimed for minimum power *with no timing constraints*. Under FF we give the number of flip-flops in the pipelined circuit. The first four circuits are 16- and 32-bit adders (carry-look-ahead, ripple-carry and carry-bypass) and the last three are multipliers. These are all 1-pipeline circuits.

We were able to achieve significant reductions in power for some of the circuits by a judicious placement of registers using the strategies described in Section 6.2. However, the maximum delay of some of the retimed circuits for low power is close to the delay of the corresponding un-pipelined circuit. Retiming for low power disregarding timing might give poor results is terms of performance.

In Table 6.2 we present the results obtained for the same circuits but now adding the constraint of minimum delay. We give results for multi-stage pipelines. The latter was obtained by applying the algorithm of Section 6.2 first to the original circuit and then to each of the two combinational parts of the retimed circuit.

We first note that the power dissipated by the pipelined circuits

Circuit Name	ST	Delay	Retime-Delay		Retime-Power		
			FF	POWER	FF	POWER	% Red
cla_16	1	12	48	2389	44	2181	8.7
	3	6	131	4632	126	4280	7.6
rpl_16	1	18	33	2303	31	2039	11.4
	3	9	98	4025	99	3698	8.1
cbp_16	1	22	38	2748	32	2407	12.4
	3	11	115	4569	105	4125	9.7
cbp_32	1	42	74	5590	59	4871	12.9
	3	21	223	9234	172	7725	16.3
mult4	1	5	14	900	13	860	4.4
	3	3	43	1503	38	1378	8.3
mult6	1	8	29	2803	26	2660	5.1
	3	4	76	3581	78	3563	0.5
mult8	1	11	46	6104	43	6003	1.7
	3	6	136	7404	128	6975	5.8

Table 6.2: Results of retiming for low power and minimum delay.

obtained by retiming for low power disregarding timing (Table 6.1) or by retiming for low power with a minimum delay constraint (Table 6.2) are very close. Thus it is possible to achieve important gains in power dissipation without loss of performance. For example rpl_16, adding the delay constraint actually results in a slightly better power dissipation. This is due to the heuristic nature of the algorithms used.

Secondly observe that, even though we are using an iterative strategy for the 3-stage pipelined circuits, the gain in power is greater for these circuits. This means that even greater savings can be obtained if the algorithm is extended to build k-stage pipelines in one pass, by taking into account in the cost function of a vertex the reduction of glitching caused by the selection of another vertex that shares a common path.

6.4 Conclusions

We described an optimization technique for low power based on retiming that is applicable to pipelined circuits. We made use of the observation that the output of registers have significantly fewer transitions than the register inputs. In particular, no glitching is present. The registers in the circuit are repositioned such that the switched capacitance $\sum_i C_i N_i$ is minimized. The results presented in Section 6.3 show that up to 16% power savings can be obtained.

The retiming algorithm for low power presented is limited to 1-stage pipelines. k-stage pipelines can be handled by iteratively applying the algorithm to the combinational logic blocks obtained after each retiming. The reason behind this limitation is that if we consider two registers in the same path, the register that is first in that path changes the switching activity on all the vertices in its transitive fanout, thus invalidating any data we have to place the second register. Further, it is very expensive to recompute the new switching activity every time the first register is moved. The solution we would obtain from an algorithm that is able to handle k-pipelines would be better than what we achieve with the current iterative approach.

For this same reason we are only considering acyclic sequential circuits. Predicting the switching activity after retiming a register in a cyclic circuit is a very difficult task. We are currently studying approximate schemes to efficiently perform this prediction and thus be able to handle both k-pipelines and cyclic sequential circuits, e.g., finite state machines.

The retiming method presented in this chapter targets the reduction of, and thus its power savings are limited by, the amount of power dissipation related to the glitching in the circuit. In the next chapters we present more powerful techniques in the sense that these techniques also reduce the power of the zero-delay switched capacitance.

References

[1] K. Lalgudi and M. Papaefthymiou. DelaY: An Efficient Tool for Retiming with Realistic Delay Modeling. In *Proceedings of the 32nd Design Automation Conference*, pages 304–309, July 1995.

[2] K. Lalgudi and M. Papaefthymiou. Fixed-Phase Retiming for Low Power Design. In *Proceedings of the International Symposium on Low Power Electronics*, pages 259–264, August 1996.

[3] C. Leiserson and J. Saxe. Optimizing Synchronous Systems. *Journal of VLSI and Computer Systems*, 1(1):41–67, Spring 1983.

[4] S. Malik, E. Sentovich, R. Brayton, and A. Sangiovanni-Vincentelli. Retiming and Resynthesis: Optimizing Sequential Networks with Combinational Techniques. *IEEE Transactions on Computer-Aided Design*, 10(1):74–84, January 1991.

[5] G. De Micheli. Synchronous Logic Synthesis: Algorithms for Cycle-Time Minimization. *IEEE Transactions on Computer-Aided Design*, 10(1):63–73, January 1991.

[6] J. Monteiro, S. Devadas, and A. Ghosh. Retiming Sequential Circuits for Low Power. In *Proceedings of the International Conference on Computer-Aided Design*, pages 398–402, November 1993.

Chapter 7

Precomputation

Power shut-down techniques, where entire modules in the circuit are "turned off" when not in use, can have a very high impact in reducing the power consumption of a circuit (cf. Section 5.3.3). We present a powerful logic optimization method that achieves data-dependent power down at the sequential or combinational logic levels.

This method is based on selectively *precomputing* the output logic values of the circuit one clock cycle before they are required, and using the precomputed values to reduce internal switching activity in the succeeding clock cycle.

The primary optimization step is the synthesis of the precomputation logic, which computes the output values for a *subset* of input conditions. If the output values can be precomputed, the original logic circuit can be "turned off" in the next clock cycle and will not have any switching activity. Since the savings in the power dissipation of the original circuit is offset by the power dissipated in the precomputation phase, the selection of the subset of input conditions for which the output is precomputed is critical. The precomputation logic adds to the circuit area and can also result in an increased clock period. Given a logic-level circuit, we present automatic methods of synthesizing the precomputation logic so as to achieve a maximal reduction in power dissipation.

We present results for two precomputation architectures for sequential circuits. The first architecture is termed *Subset Input Disabling* architecture [1] and is described in Section 7.1. In this architecture the precomputation logic is determined from a subset of the primary inputs to the original circuit. For the second sequential precomputation architecture, the *Complete Input*

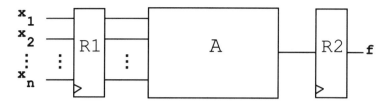

Figure 7.1: Original circuit.

Disabling architecture [6] of Section 7.2, the precomputation logic can be a function of all the input variables. The complete input disabling architecture can reduce power dissipation for a larger class of sequential circuits than the subset input disabling architecture, but the synthesis of the precomputation logic block is more complex.

We extend the precomputation approach to combinational logic circuits [6]. The reduction in switching activity is achieved by introducing transmission-gates or transparent latches in the circuit which can be disabled when the signal going through them is not necessary to determine the output values. This architecture is more flexible than any of the sequential architectures since we are not limited to precomputation over primary inputs. However, these degrees of freedom make the optimization step much harder. We present synthesis methods for precomputation of combinational circuits. Synthesis methods that target this combinational architecture as well as other variants have been independently developed in [9].

For each of these precomputation architectures, we present experimental results that show that power savings up to 75 percent can be achieved.

7.1 Subset Input Disabling Precomputation

Consider the circuit of Figure 7.1. We have a combinational logic block A that is bounded by registers R_1 and R_2. While R_1 and R_2 are shown as distinct set of registers in Figure 7.1 they could, in fact, be the same registers. We will first assume that block A has a single output and that it implements the Boolean function f.

7. PRECOMPUTATION

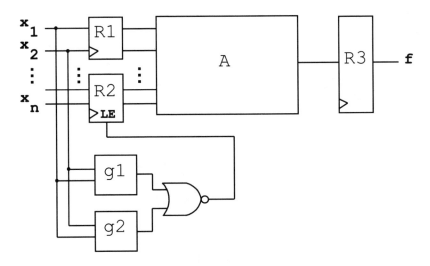

Figure 7.2: Subset input disabling precomputation architecture.

7.1.1 Subset Input Disabling Precomputation Architecture

In Figure 7.2 we show the *Subset Input Disabling* precomputation architecture. The inputs to the block A have been partitioned into two sets, corresponding to the registers R_1 and R_2. The output of the logic block A feeds the register R_3. The two Boolean functions g_1 and g_2 are the *predictor* functions. We require:

$$g_1 = 1 \Rightarrow f = 1 \tag{7.1}$$

$$g_2 = 1 \Rightarrow f = 0 \tag{7.2}$$

g_1 and g_2 only depend on the subset of the inputs to f going into R_1. If g_1 or g_2 evaluates to a 1 during clock cycle t, the load enable signal to the register R_2 is turned off. This implies that the outputs of R_2 during clock cycle $t+1$ do not change. However, the outputs of register R_1 are updated and g_1 or g_2 evaluating to 1 indicate that the subset of inputs feeding R_1 are enough to compute f, hence the function f will evaluate to the correct logical value.

A power reduction is achieved because only a subset of the inputs to block A change implying reduced switching activity. Though, the area of the circuit has increased due to additional logic corresponding to g_1, g_2 and the NOR gate. The delay of the circuit between R_1/R_2 and R_3 is unchanged. However, g_1 and g_2 add to the delay of paths that originally ended at R_1 but now pass through g_1 or g_2 and the NOR gate before ending at the load enable

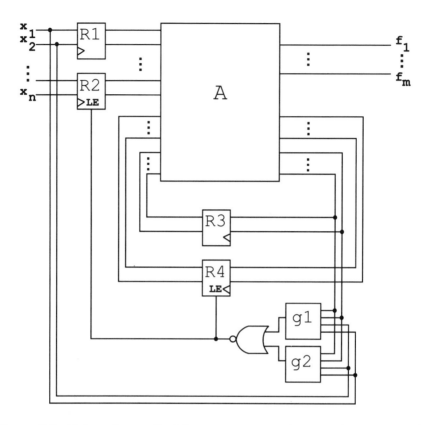

Figure 7.3: Subset input disabling precomputation architecture applied to a finite state machine.

signal of the register R_2. Therefore, we would like to apply this transformation on non-critical logic blocks or choose the input signals to the precomputation such that they are not in the critical path.

The choice of g_1 and g_2 is critical. We wish to include as many input conditions as we can in g_1 and g_2. In other words, we wish to maximize the probability of g_1 or g_2 evaluating to a 1. In the extreme case, this probability can be made unity if $g_1 = f$ and $g_2 = \bar{f}$. However, this would imply a duplication of the logic block A and no reduction in power with a twofold increase in area! To obtain reduction in power with marginal increases in circuit area and delay, g_1 and g_2 have to be significantly less complex than f. One way of ensuring this is to make g_1 and g_2 depend on much fewer inputs than f.

As mentioned before, the sequential precomputation architectures are not restricted to pipeline circuits. We present in Figure 7.3 an example

7. PRECOMPUTATION

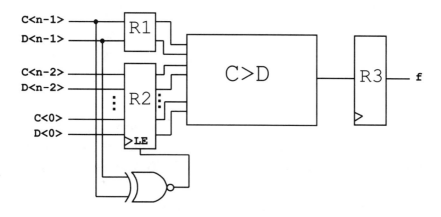

Figure 7.4: A comparator example.

of precomputation for a finite state machine using this subset input disabling precomputation architecture.

7.1.2 An Example

We give an example that illustrates the fact that substantial power gains can be achieved with marginal increases in circuit area and delay. The circuit we are considering is a n-bit comparator that compares two n-bit numbers C and D and computes the function $C > D$. The optimized circuit with precomputation logic is shown in Figure 7.4. The precomputation logic is as follows.

$$g_1 = C\langle n-1 \rangle \cdot \overline{D\langle n-1 \rangle}$$
$$g_2 = \overline{C\langle n-1 \rangle} \cdot D\langle n-1 \rangle$$

Clearly, when $g_1 = 1$, C is greater than D, and when $g_2 = 1$, C is less than D. We have to implement

$$\overline{g_1 + g_2} = C\langle n-1 \rangle \oplus D\langle n-1 \rangle$$

where $\oplus$ stands for the exclusive-or operator.

Assuming a uniform probability for the inputs, i.e., $C\langle i \rangle$ and $D\langle i \rangle$ have a 0.5 static probability of being a 0 or a 1, the probability that the XOR gate evaluates to a 1 is 0.5, regardless of n. For large n, we can neglect the power dissipation in the XOR gate, and therefore, we can achieve a power reduction close to 50%. The reduction will depend upon the relative power dissipated by the vector pairs with $C\langle n-1\rangle \oplus D\langle n-1\rangle = 1$ and the vector pairs with $C\langle n-1\rangle \oplus D\langle n-1\rangle = 0$. If we add the inputs $C\langle n-2\rangle$ and $D\langle n-2\rangle$ to g_1 and g_2 it is possible to achieve a power reduction close to 75%.

7.1.3 Synthesis of Precomputation Logic

In this section, we describe exact and approximate methods to determine the functionality of the precomputation logic for the subset input disabling architecture, and then describe methods to efficiently implement the logic.

Precomputation and Observability Don't-Cares

Assume that we have a logic function $f(X)$ corresponding to block A of Figure 7.1, with $X = \{x_1, \cdots, x_n\}$. Given that the logic function implemented by block A is f, then the *Observability Don't-Care Set* (cf. Section 5.2.2) for input x_i is given by:

$$ODC_i = f_{x_i} \cdot f_{\overline{x_i}} + \overline{f}_{x_i} \cdot \overline{f}_{\overline{x_i}} \qquad (7.3)$$

where f_{x_i} and $f_{\overline{x_i}}$ are the cofactors of f with respect to x_i, and similarly for $\overline{f}$.

If we determine that a given input combination is in ODC_i then we can disable the loading of x_i into the register since that means that we do not need the value of x_i in order to know what the value of f is. If we wish to disable the loading of registers $x_m, x_{m+1}, \cdots, x_n$, we will have to implement the function

$$g = \prod_{i=m}^{n} ODC_i \qquad (7.4)$$

and use $\overline{g}$ as the load enable signal for the registers corresponding to $x_m, x_{m+1}, \cdots, x_n$.

Precomputation Logic

Let us now consider the subset input disabling architecture of Figure 7.2. Assume that the inputs $x_1, \cdots, x_m$, with $m < n$ have been selected as the variables that g_1 and g_2 depend on. We have to find g_1 and g_2 such that they satisfy the constraints of Equations 7.1 and 7.2, respectively, and such that $prob(g_1 + g_2)$ is maximum.

7. PRECOMPUTATION

We can determine g_1 and g_2 using universal quantification on f. The *universal quantification* of a function f with respect to a variable x_i is defined as:

$$U_{x_i} f = f_{x_i} \cdot f_{\overline{x_i}} \tag{7.5}$$

This gives all the combinations over the inputs $x_1, \cdots, x_{i-1}, x_{i+1}, \cdots, x_n$, that result in $f = 1$ independently of the values of x_i.

Given a subset of inputs $S = \{x_1, \cdots, x_m\}$, let $D = X - S$. We can define:

$$U_D f = U_{x_{m+1}} \ldots U_{x_n} f \tag{7.6}$$

Theorem 7.1 $g_1 = U_D f$ *satisfies Equation 7.1. Further, no other function* $h(x_1, \cdots, x_m)$ *exists such that* $prob(h) > prob(g_1)$ *and such that* $h = 1 \Rightarrow f = 1$.

Proof—If, for some input combination $a_1, \cdots, a_m$, $g_1(a_1, \cdots, a_m) = 1$, then by construction for that combination of $x_1, \cdots, x_m$ and all possible combinations of variables in $x_{m+1}, \cdots, x_n$, $f(a_1, \cdots, a_m, x_{m+1}, \cdots, x_n) = 1$.

We cannot add any minterm over $x_1, \cdots, x_m$ to g_1 because for any minterm that is added, there will be some combination of $x_{m+1}, \cdots, x_n$ for which $f(x_1, \cdots, x_n)$ will evaluate to a 0. Therefore, we cannot find any function h that satisfies Equation 7.1 and such that $prob(h) > prob(g_1)$. ∎

Similarly, given a subset of inputs S, we can obtain a maximal g_2 by:

$$g_2 = U_D \overline{f} = U_{x_{m+1}} \ldots U_{x_n} \overline{f} \tag{7.7}$$

We can compute the functionality of the precomputation logic as $g_1 + g_2$.

Selecting a Subset of Inputs: Exact Method

Given a function f we wish to select the "best" subset of inputs S of cardinality k. Given S, we have $D = X - S$ and we compute $g_1 = U_D f$, $g_2 = U_D \overline{f}$. In the sequel, we assume that the best set of inputs corresponds to the inputs which result in $prob(g_1 + g_2)$ being maximum for a given k. We know that $prob(g_1 + g_2) = prob(g_1) + prob(g_2)$ since g_1 and g_2 cannot both be 1 for the same input vector. The above cost function ignores the power dissipated in the precomputation logic, but since the number of inputs to the precomputation logic is significantly smaller than the total number of inputs this is a good approximation.

```
1.      SELECT_INPUTS( f, k ):
2.      {
3.          /* f = function to precompute */
4.          /* k = # of inputs to precompute with */
5.          BEST_PROB = 0 ;
6.          SELECTED_SET = φ ;
7.          SELECT_RECUR( f, f̄, φ, X, |X| − k ) ;
8.          return( SELECTED_SET ) ;
9.      }
10.
11.     SELECT_RECUR( $f_a, f_b, D, Q, l$ ):
12.     {
13.         if( |D| + |Q| < l )
14.             return ;
15.         pr = prob($f_a$ = 1) + prob($f_b$ = 1) ;
16.         if( pr ≤ BEST_IN_PROB )
17.             return ;
18.         else if( |D| == l ) {
19.             BEST_IN_PROB = pr ;
20.             SELECTED_SET = X − D ;
21.             return ;
22.         }
23.         select next $x_i ∈ Q$ ;
24.         SELECT_RECUR( $U_{x_i}f_a, U_{x_i}f_b, D ∪ x_i, Q − x_i, l$ ) ;
25.         SELECT_RECUR( $f_a, f_b, D, Q − x_i, l$ ) ;
26.     }
```

Figure 7.5: Procedure to determine the optimal subset of inputs to the precomputation logic.

We describe a branching algorithm that determines the optimal set of inputs maximizing the probability of the g_1 and g_2 functions. This algorithm is shown in pseudo-code in Figure 7.5.

The procedure **SELECT_INPUTS** receives as arguments the function f and the desired number of inputs k to the precomputation logic. SE-

LECT_INPUTS calls the recursive procedure **SELECT_RECUR** with five arguments. The first two arguments correspond to the g_1 and g_2 functions, which are initially f and $\bar{f}$. A variable is selected within the recursive procedure and the two functions are universally quantified with respect to the selected variable. The third argument D corresponds to the set of variables that g_1 and g_2 do *not* depend on. The fourth argument Q corresponds to the set of "active" variables, which may still be selected or discarded. Finally, the argument l corresponds to the number of variables that have to be universally quantified in order to obtain g_1 and g_2 with k or fewer inputs.

If the condition of line 13 ($|D|+|Q| < l$) is true then we have dropped too many variables in the earlier recursions and we will not be able to quantify with respect to enough input variables. The functions g_1 and g_2 will depend on too many variables ($> k$).

We calculate the probability of $g_1 + g_2$ (line 15). If this probability is less than the maximum probability we have encountered thus far, we can immediately return since the following invariant

$$prob(U_x f) = prob(f_{x_i} \cdot f_{\bar{x_i}}) \leq prob(f) \quad \forall x_i, f \qquad (7.8)$$

is true because f contains $U_x f$. Therefore as we universally quantify variables from a given f_a and f_b function pair, the *pr* quantity monotonically decreases.

We store the selected set corresponding to the maximum probability found.

Selecting a Subset of Inputs: Approximate Method

The worst-case running time of the exact method is exponential in the number of input variables and although we have a nice pruning condition, there are many examples for which we cannot apply this method. Thus we have also implemented an approximate algorithm that looks at each primary input individually and chooses the k most promising inputs.

For each input we calculate:

$$p_i = prob(U_x f) + prob(U_x \bar{f}) \qquad (7.9)$$

p_i is the probability that we know the value of f without knowing the value of x_i. If p_i is high then most of the time we do not need x_i to compute f. We select the k inputs corresponding to smaller values of p_i.

Implementing the Logic

The Boolean operations of OR and universal quantification required in the input selection procedure can be carried out efficiently using reduced, ordered Binary Decision Diagrams (ROBDDs) [4]. We obtain a ROBDD for the $g_1 + g_2$ function. A ROBDD can be converted into a multiplexor-based network (see [2]) or into a sum-of-products cover. The network or cover can then be optimized using standard combinational logic optimization methods that reduce area [3] or those that target low power dissipation [8].

7.1.4 Multiple-Output Functions

In general, we have a multiple-output function $f_1, \cdots, f_m$ that corresponds to the logic block A in Figure 7.1. All the procedures described thus far can be generalized to the multiple-output case.

The functions g_{1i} and g_{2i} are obtained using the equations below.

$$g_{1i} = U_D f_i \qquad (7.10)$$
$$g_{2i} = U_D \overline{f_i} \qquad (7.11)$$

where $D = X - S$ is given as before. The function g whose complement drives the load enable signal is obtained as:

$$g = \prod_{i=1}^{m}(g_{1i} + g_{2i}) \qquad (7.12)$$

The function g corresponds to the set of input conditions where the variables in S control the values of *all* the f_i's regardless of the values of variables in $D = X - S$.

Selecting a Subset of Outputs: Exact Method

The probability that g, as defined in Equation 7.12, is 1 may be very low since the number input combinations that allow precomputation of all outputs may be very small. We describe an algorithm, which given a multiple-output function, selects a subset of outputs *and* a subset of inputs so as to maximize a given cost function that is dependent on the probability of the precomputation logic and the number of selected outputs. This algorithm is described in the pseudo-code of Figure 7.6.

7. PRECOMPUTATION

```
1.     SELECT_OUTPUTS( F = {f₁, ···, fₘ}, k ):
2.     {
3.         /* F = multiple-output function to precompute */
4.         /* k = # of inputs to precompute with */
5.         BEST_OUT_COST = 0 ;
6.         SEL_OP_SET = φ ;
7.         SELECT_ORECUR( φ, F, 1, k ) ;
8.         return( SEL_OP_SET ) ;
9.     }
10.
11.    SELECT_ORECUR( G, H, proldG, k ):
12.    {
13.        lf = gates(G ∪ H)/total_gates × proldG ;
14.        if( lf ≤ BEST_OUT_COST )
15.            return ;
16.        if( G ≠ φ )
17.            if( SELECT_INPUTS( G, k ) == φ )
18.                return ;
19.        prG = BEST_IN_PROB ;
20.        cost = prG × gates(F − H)/total_gates ;
21.        if( cost > BEST_OUT_COST) {
22.            BEST_OUT_COST = cost ;
23.            SEL_OP_SET = G ;
24.        }
25.        select next fᵢ ∈ H ;
26.        SELECT_ORECUR( G ∪ fᵢ, H − fᵢ, prG, k ) ;
27.        SELECT_ORECUR( G, H − fᵢ, prG, k ) ;
28.    }
```

Figure 7.6: Procedure to determine the optimal set of outputs.

The inputs to procedure **SELECT_OUTPUTS** are the multiple-output function F, and a number k corresponding to the number of inputs to the precomputation logic.

The procedure **SELECT_ORECUR** receives as inputs two sets G

and H, which correspond to the current set of outputs that have been selected and the set of outputs which can be added to the selected set, respectively. Initially, $G = \phi$ and $H = F$. The cost of a particular selection of outputs, namely G, is given by $prG \times$ gates$(F - H)$/total_gates, where prG corresponds to the signal probability of the precomputation logic, gates$(F - H)$ corresponds to the number of gates in the logic corresponding to the outputs in G and not shared by any output in H, and total_gates corresponds to the total number of gates in the network (across all outputs of F).

There are two pruning conditions that are checked for in the procedure **SELECT_ORECUR**. The first corresponds to assuming that all the outputs in H can be added to G without decreasing the probability of the precomputation logic. This is a valid condition because the quantity *proldG* in each recursive call can only decrease with the addition of outputs to G. The second condition is that to be able to precompute G we may need variables already discarded. Therefore prG will always be 0 for lower recursion levels.

Logic Duplication

Since we are only precomputing a subset of outputs, we may incorrectly evaluate the outputs that we are *not* precomputing as we disable certain inputs during particular clock cycles. If an output that is not being precomputed depends on an input that is being disabled, then the output will be incorrect.

The support of f, denoted as *support*(f), is the set of all variables x_i that occur in f as x_i or $\overline{x_i}$. Once a set of outputs $G \subset F$ and a set of precomputation logic inputs $S \subset X$ have been selected, we need to duplicate the registers corresponding to $(support(G) - S) \cap support(F - G)$. The inputs that are being disabled are in $support(G) - S$. Logic in the $F - G$ outputs that depends on the set of duplicated inputs has to be duplicated as well. It is precisely for this reason that we maximize $prG \times$ gates$(F - H)$ rather than $prG \times$ gates(G) in the output-selection algorithm. This way we are maximizing the number of gates (logic corresponding to the outputs in G) that will not switch when precomputation is possible but not taking into account gates that are shared by the outputs in H, thus reducing the amount of duplication as much as possible.

An example of a multiple-output function where registers and logic need to be duplicated is shown in Figure 7.7.

The original network of Figure 7.7(a) has outputs f_1 and f_2 and inputs

7. PRECOMPUTATION

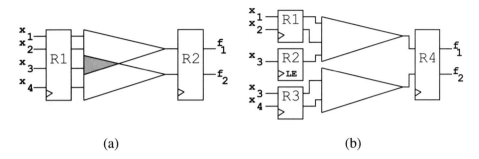

Figure 7.7: Logic duplication in a multiple-output function.

$x_1, \cdots, x_4$. The function f_1 depends on inputs x_1, x_2 and x_3 and the function f_2 depends on inputs x_3 and x_4. Hence, the two outputs are sharing the input x_3. Suppose that the output-selection procedure determines that f_1 is the best output to precompute and that inputs x_1 and x_2 are the best inputs to the precomputation logic. Therefore, just as in the case of a single-output function, the inputs x_1 and x_2 feed the input register, whereas x_3 feeds the register with the load-enable signal. However, since f_2 depends on x_3 and the register with the load-enable signal contains stale values in some clock cycles. We need to duplicate the register for x_3 and the logic from x_3 to f_2.

Selecting a Subset of Outputs: Approximate Method

Again the exact algorithm for output selection is worst-case exponential in the number of inputs plus number of outputs, thus we need an approximate method to handle larger circuits. We designed an approximate algorithm which is presented in pseudo-code in Figure 7.8.

In this algorithm we first select the set of outputs that will be precomputed and then select the inputs that we are going to precompute those outputs with. When we are selecting the outputs we still do not know which inputs are going to be selected, thus we select those outputs that seem to be the *most precomputable*. Universally quantifying just one of the inputs, we start with one output and compute the same cost function as in the exact method, $prG \times$ gates$(F - H)$/total_gates. Then we add outputs that make the cost function increase. We repeat this process for each input. At the end we keep the set of outputs corresponding to the maximum cost.

Once we have a set of promising outputs to precompute we can use

```
1.   SELECT_OUTPUTS_APPROX( F = {f_1, ···, f_m}, k ):
2.   {
3.       BEST_OUT_COST = 0 ;
4.       foreach x_i ∈ X {         /* Output selection */
5.           foreach f_j ∈ F
6.               g_j = U_x f_j + U_x \overline{f_j} ;
7.           foreach f_j ∈ F {
8.               G = {f_j} ;
9.               H = F − {f_j} ;
10.              prG = prob(g_j) ;
11.              curr_cost = prG × gates(F − H)/total_gates;
12.              /* Add any outputs that make the cost increase */
13.              g = g_j ;
14.              foreach f_l ∈ F {
15.                  H = H − {f_l} ;
16.                  prG = prob(g · g_l) ;
17.                  cost = prG × gates(F − H)/total_gates ;
18.                  if( cost > curr_cost ) {
19.                      curr_cost = cost ;
20.                      g = g · g_l ;
21.                      G = G ∪ {f_l} ;
22.                  } else
23.                      H = H ∪ {f_l} ;
24.              }
25.          }
26.          if( curr_cost > BEST_OUT_COST ) {
27.              BEST_OUT_COST = curr_cost ;
28.              SEL_OP_SET = G ;
29.          }
30.      }
31.      foreach x_i ∈ X {         /* Input selection */
32.          g = 1 ;
33.          foreach f_j ∈ SEL_OP_SET
34.              g = g · (U_x f_j + U_x \overline{f_j}) ;
35.          p_i = prob(g) ;
36.      }
37.      select k x_i's, corresponding to smaller p_i's
38.  }
```

Figure 7.8: Procedure to determine a good subset of outputs.

7. PRECOMPUTATION

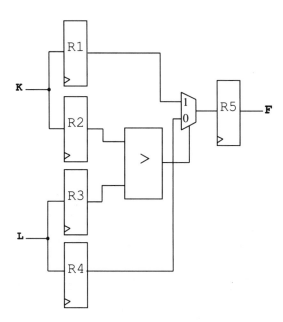

Figure 7.9: Precomputation applied to a maximum circuit.

the approximate algorithm described in Section 7.1.3 to select the inputs. This algorithm runs in polynomial time in the number outputs times the number of inputs.

7.1.5 Examples of Precomputation Applied to Datapath Modules

Some datapath modules are particularly well suited for the subset input disable precomputation architecture. An example of this are n-bit comparators, as the one depicted in Figure 7.4. We give examples of two other such circuits.

A *MAX* function can be implement as shown in Figure 7.9. The input registers are duplicated so that we can perform precomputation on the comparator just like in Figure 7.4, where some of the inputs of R_2 and R_3 are disabled. Further, the enable signal from the precomputation logic can be used to only enable either R_1 or R_4.

Another datapath module for which significant power savings can be achieved with this sequential precomputation architecture is a carry-select adder, shown in Figure 7.10. In order to reduce the time per operation, the addition of the most significant bits $A\langle 8 : 15\rangle$ and $B\langle 8 : 15\rangle$ is done in parallel

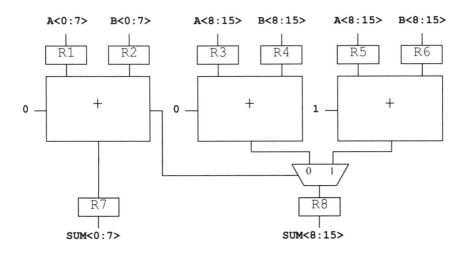

Figure 7.10: Precomputation applied to a carry-select adder.

for the two cases where there is a carry from the addition of $A\langle 0:7\rangle$ and $B\langle 0:7\rangle$ or there is no carry.

We can make

$$g_1 = A\langle 7\rangle \cdot B\langle 7\rangle$$

be the latch enable for registers R_3 and R_4 as in this case we know there is going to be a carry. Similarly

$$g_2 = \overline{A\langle 7\rangle} \cdot \overline{B\langle 7\rangle}$$

can be used as the latch enable for R_5 and R_6. Using this scheme, we will be eliminating all the switching activity in one of the adders of Figure 7.10 for half of the input combinations, corresponding to approximately 16% power savings.

7.1.6 Multiple Cycle Precomputation

Basic Strategy

It is possible to precompute output values that are not required in the succeeding clock cycle, but required 2 or more clock cycles later.

Consider the topology of Figure 7.11. If the outputs of register R_3 are not used except to compute f, then we can precompute the value of the function f using a selected set of inputs, namely those corresponding to register

7. PRECOMPUTATION

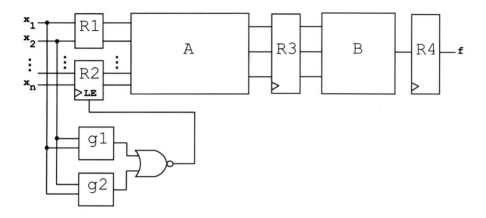

Figure 7.11: Multiple cycle precomputation.

R_1. If f can be precomputed to a 1 or a 0 for a set of input conditions, then for these inputs we can turn off the load enable signal to R_2. This will reduce switching activity not only in logic block A, but also in logic block B, because there will be reduced switching activity at the outputs of R_3 in the clock cycle following the one where the outputs of R_2 do not change.

Examples

We present some examples illustrating multiple-cycle precomputation.

Consider the circuit of Figure 7.12. The function f computes $(C + D) > (X+Y)$ in two clock cycles. Attempting to precompute $C+D$ or $X+Y$ using the methods of the previous sections does not result in any savings because there are too many outputs to consider. However, 2-cycle precomputation can reduce switching activity by close to 12.5% if the functions below are used.

$$g_1 = C\langle n-1\rangle \cdot D\langle n-1\rangle \cdot \overline{X\langle n-1\rangle} \cdot \overline{Y\langle n-1\rangle}$$
$$g_2 = \overline{C\langle n-1\rangle} \cdot \overline{D\langle n-1\rangle} \cdot X\langle n-1\rangle \cdot Y\langle n-1\rangle$$

where g_1 and g_2 satisfy the constraints of Equations 7.1 and 7.2, respectively. Since $prob(g_1 + g_2) = \frac{2}{16} = 0.125$, we can disable the loading of registers $C\langle n-2:0\rangle$, $D\langle n-2:0\rangle$, $X\langle n-2:0\rangle$, and $Y\langle n-2:0\rangle$ 12.5% of the time, which results in switching activity reduction. This percentage can be increased to over 45% by using $C\langle n-2\rangle$ through $Y\langle n-2\rangle$. We can additionally

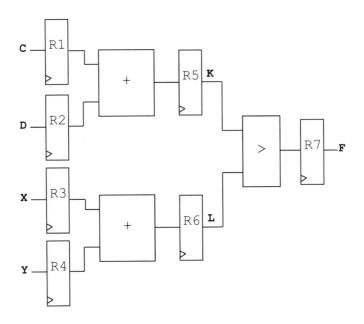

Figure 7.12: Adder-comparator circuit.

use single-cycle precomputation logic (as illustrated in Figure 7.4) to further reduce switching activity in the > comparator of Figure 7.12.

Next, consider the circuit of Figure 7.13. The multiple-output function f computes $MAX(C + D, X + Y)$ in two clock cycles. We can use exactly the same g_1 and g_2 functions as those immediately above, but g_1 is used to disable the loading of registers $X\langle n - 2 : 0\rangle$ and $Y\langle n - 2 : 0\rangle$, and g_2 is used to disable the loading of $C\langle n - 2 : 0\rangle$ and $D\langle n - 2 : 0\rangle$. We exploit the fact that if we know that $C + D > X + Y$, there is no need to compute $X + Y$, and vice versa.

7.1.7 Experimental Results for the Subset Input Disabling Architecture

We first present in Table 7.1 results on datapath circuits such as carry-select adders, comparators, and interconnections of adders and comparators. In all examples all the outputs of each circuit were precomputed. For each circuit, we give the number of literals (LTS), levels of logic (L) and power (P) of the original circuit under ORIGINAL, the number of inputs (I), literals (LTS) and levels (L) of the precomputation logic under PRECOMP. LOGIC, the final power

7. PRECOMPUTATION

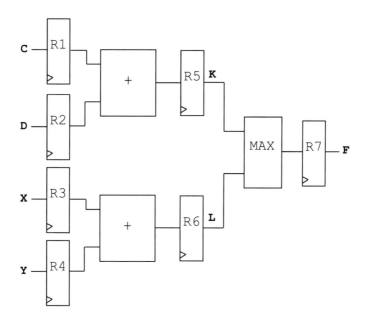

Figure 7.13: Adder-maximum circuit.

(P) and the percent reduction in power (% RED) under OPTIMIZED. All power estimates are in micro-Watt and are computed using the techniques described in Chapter 4. A zero delay model, a clock frequency of 20MHz and a supply voltage of 5V were assumed. The rugged script of SIS [7] was used to optimize the precomputation logic.

Power dissipation decreases in almost all cases. For circuit comp16, a 16-bit parallel comparator, the power savings increase as more inputs are used in the precomputation logic, up to 60% when 8 inputs are used for precomputation. When 10 inputs are used, the savings go down to 58% as the size of the precomputation logic offsets the larger amount of time that we are disabling the other input registers.

Results for multiple-cycle precomputation are given for circuits add_comp16 and add_max16, shown in Figures 7.12 and 7.13 respectively. For circuit add_comp16, for instance, the numbers 4/8 under the fifth column indicate that 4 inputs are used to precompute the adders in the first cycle and 8 inputs are used to precompute the comparator in the next cycle.

The number of levels of the precomputation logic is an indication of the performance penalty in using precomputation. The logic that is driving the input flip-flops to the original circuit is increased in depth by the number of

7.1. SUBSET INPUT DISABLING PRECOMPUTATION

Circuit Name	Original			Precomp. Logic			Optimized	
	LTS	L	P	I	LTS	L	P	% RED
comp16	286	7	1281	2	4	2	965	25
				4	8	2	683	47
				6	12	2	550	57
				8	16	2	518	60
				10	20	2	538	58
max16	350	9	1744	8	16	2	1281	27
csa16	975	10	2945	2	4	2	2958	0
				4	11	4	2775	6
				6	18	4	2676	9
				8	25	5	2644	10
add_comp16	3026	8	6941	4/0	8	2	6346	9
				4/8	24	4	5711	18
				8/0	51	4	4781	31
				8/8	67	6	3933	43
add_max16	3090	9	7370	4/0	8	2	7174	3
				4/8	24	4	6751	8
				8/0	51	4	6624	10
				8/8	67	6	6116	17

Table 7.1: **Power reductions for datapath circuits.**

levels of the precomputation logic. In most cases, the increase in the number of levels is small.

Results on random logic circuits are presented in Table 7.2. The random logic circuits are taken from the MCNC combinational benchmark sets. In our experiments we assumed that the inputs to the circuits are outputs of flip-flops, and applied sequential precomputation. We give results for those examples where significant savings in power was obtained.

Again, the subset input disabling precomputation architecture was used and the input and output selection algorithms described in Sections 7.1.3 and 7.1.4 were used. Due to the size of the circuits, on most examples the approximate algorithm was used. Circuits for which we were able to run the exact algorithm are marked with a *. The columns in this table have the same meaning as in Table 7.1, except for the second and third columns which show the number of inputs (I) and outputs (O) of each circuit, and the eighth column which shows the number of outputs that are being precomputed (O). It is noteworthy that in some cases, as much as 75% reduction in power dissipation is obtained.

7. PRECOMPUTATION

Circuit Name	Original					Precompute Logic				Optimized	
	I	O	LTS	L	P	I	O	LTS	L	P	RED
apex2	39	3	395	11	2387	4	3	4	1	1378	42
cht	47	36	167	3	1835	1	35	1	1	1537	16
cm138*	6	8	35	2	286	3	8	3	1	153	47
cm150*	21	1	61	4	744	1	1	1	1	574	23
cmb*	16	4	62	5	620	5	4	10	1	353	43
comp	32	3	185	6	1352	6	3	13	2	627	54
cordic*	23	2	194	13	1049	10	2	18	2	645	39
cps	24	109	1203	9	3726	7	101	26	3	2191	41
dalu	75	16	3067	24	11048	5	16	12	2	7344	34
duke2	22	29	424	7	1732	9	29	24	3	1328	23
e64	65	65	253	32	2039	5	65	5	1	513	75
i2	201	1	230	3	5606	17	1	42	5	1943	65
major*	5	1	12	3	173	1	1	1	1	141	19
misex2	25	18	113	5	976	8	18	16	3	828	15
misex3	25	18	626	14	2350	2	14	2	1	1903	19
mux*	21	1	54	5	715	1	1	0	0	557	22
pcle	19	9	71	7	692	3	9	3	1	486	30
pcler8	27	17	95	8	917	3	17	3	1	571	38
sao2*	10	4	270	17	1191	2	4	2	1	422	65
seq	42	35	1724	11	6112	2	35	1	1	2134	65
spla	16	46	634	9	2267	4	46	6	1	1340	41
term1	34	10	625	9	3605	8	10	14	3	2133	41
toolarge	38	3	491	11	2718	1	3	1	1	1756	35
unreg	36	16	144	2	1499	2	15	2	1	1234	18

*Precomputation logic calculated using the exact algorithm.

Table 7.2: Power reductions for random logic circuits.

The area penalty incurred is indicated by the number of literals in the precomputation logic and is 3% on the average. The extra delay incurred is proportional to the number of levels in the precomputation logic and is quite small in most cases.

7.2 Complete Input Disabling Precomputation

The precomputation architecture presented in the previous section suffers from the limitation that if a logic function is dependent on the values of several inputs for a large fraction of the applied input combinations, then

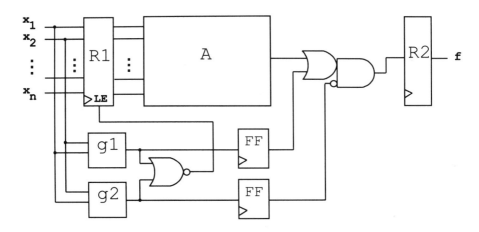

Figure 7.14: Complete input disabling precomputation architecture.

no reduction in switching activity can be obtained since we cannot build the precomputation logic from any small subset of the primary inputs.

In this section we target a general precomputation architecture, termed *Complete Input Disabling*, for sequential logic circuits and show that it is significantly more powerful than the subset input disabling architecture previously described. The very power of this architecture makes the synthesis of precomputation logic a challenging problem. We present a method to automatically synthesize precomputation logic for this architecture.

7.2.1 Complete Input Disabling Precomputation Architecture

In Figure 7.14 the second precomputation architecture for sequential circuits is shown. We are again assuming that the original circuit is in the form of Figure 7.1. However, the complete input disabling architecture is also applicable to cyclic sequential circuits. The functions g_1 and g_2 satisfy the conditions of Equations 7.1 and 7.2 as before. During clock cycle t if either g_1 or g_2 evaluates to a 1, we set the load enable signal of the register R_1 to be 0. This means that in clock cycle $t + 1$ the inputs to the combinational logic block A do not change implying zero switching activity. If g_1 evaluates to a 1 in clock cycle t, the input to register R_2 is a 1 in clock cycle $t + 1$, and if g_2 evaluates to a 1, then the input to register R_2 is a 0. Note that g_1 and g_2 cannot both be 1 during the same clock cycle due to the conditions imposed by Equations 7.1 and 7.2.

7. PRECOMPUTATION

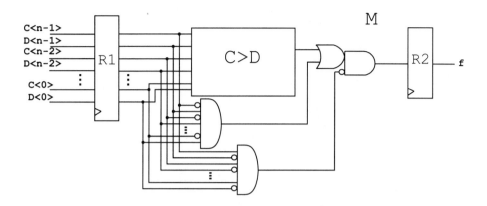

Figure 7.15: A modified comparator.

The important difference between this architecture and the subset input disabling architecture is that the precomputation logic can be a function of all input variables, allowing us to precompute any input combination. We have additional logic corresponding to the two flip-flops marked FF and the AND-OR gate shown in the figure. Also the delay between R_1 and R_2 has increased due to the addition of this gate.

Note that for all input combinations that are included in the precomputation logic (corresponding to $g_1 + g_2$) we are not going to use the output of f. Therefore we can simplify the combinational logic block A by using these input combinations as an *Observability Don't-Care Set* for f.

7.2.2 An Example

A simple example that illustrates the effectiveness of the subset input disabling architecture is a *n*-bit comparator. The precomputed comparator is shown in Figure 7.4.

Now let us consider a modified comparator, as shown in Figure 7.15. It works just like a *n*-bit comparator except that if C is equal to the all 0's bit-vector and D is equal to the all 1's bit-vector the result should still be 1 and vice-versa, if C is equal to the all 1's bit-vector and D is equal to the all 0's bit-vector the result should still be 0. This circuit is not precomputable using the subset input disabling architecture because knowing that $C\langle n-1\rangle = 0$ and $D\langle n-1\rangle = 1$ or $C\langle n-1\rangle = 1$ and $D\langle n-1\rangle = 0$ is not enough information to infer the value of f. In fact, we need to know the values of *all* the inputs

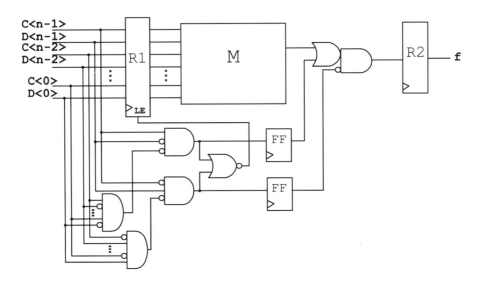

Figure 7.16: Modified comparator under the complete input disabling architecture.

in order to determine f. Thus, although the input combination C equal to the all 0's bit-vector and D equal to the all 1's, and the input combination C equal to the all 1's bit-vector and D equal to the all 0's bit-vector have a very low probability of occurrence, they invalidate the use of the subset input disabling precomputation architecture.

Using the complete input disabling architecture, since we have access to all input variables for the precomputation logic, we can simply remove these input combinations from g_2 and g_1, respectively. This is illustrated in Figure 7.16. This way we will still be precomputing all other input combinations in $C\langle n-1\rangle \oplus D\langle n-1\rangle$, meaning that the fraction of the time that we will precompute the output value is still close to 50%.

7.2.3 Synthesis of Precomputation Logic

The key tradeoff in selecting the precomputation logic is that we want to include in it as many input combinations as possible but at the same time keep this logic simple. The subset input disabling precomputation architecture ensures that the precomputation logic is significantly less complex than the combinational logic block A in the original circuit by restricting the search space to identifying g_1 and g_2 such that they depend on a relatively small

7. PRECOMPUTATION

subset of the inputs to A.

By making the precomputation logic depend on all inputs, the complete input disabling architecture allows for a greater flexibility but also makes the problem much more complex. The algorithm to determine the precomputation logic that we present in this section extends the algorithm of Section 7.1.3 to exploit this greater flexibility.

We will be searching for the subset of inputs that, for a large fraction of the input combinations, are necessary to determine what the value of f is. We follow a strategy of keeping the precomputation logic simple by making the logic depend *mostly* on a small subset of inputs. The difference is that now we are not going to restrict ourselves to those input combinations for which this subset of inputs defines f, we will allow for some input combinations that need inputs not in the selected set.

Selecting a Subset of Inputs: Exhaustive Method

Given a function f we are going to select the "best" subset of inputs S of cardinality k such that we minimize the number of times we need to know the value of the other inputs to evaluate f. For each subset of size k, we compute the cofactors of f with respect to all combinations of inputs in the subset. If the probability of a cofactor of f with respect to a cube c is close to 1 (or close to 0), it means that for the combination of input variables in c the value of f will be 1 (or 0) most of the time.

Let us consider f with inputs $x_1, x_2, \cdots, x_n$ and assume that we have selected the subset $x_1, x_2, \cdots, x_k$. If the probability of the cofactor of f with respect to $x_1 x_2 \cdots x_k$ being all 1's is high (i.e., prob($f_{x_1 x_2 \cdots x_k}$) $\approx$ 1), then over all combinations of $x_{k+1}, \cdots, x_n$ there are only a few for which f is not 1. So we can include $x_1 x_2 \cdots x_k \cdot f_{x_1 x_2 \cdots x_k}$ in g_1. Similarly if the probability of the $f_{x_1 x_2 \cdots x_k}$ is low (i.e., prob($f_{x_1 x_2 \cdots x_k}$) $\approx$ 0), then over all combinations of $x_{k+1}, \cdots, x_n$ there are only a few for which f is not 0, so we include $x_1 x_2 \cdots x_k \cdot \overline{f}_{x_1 x_2 \cdots x_k}$ in g_2. Note that in the subset input disabling architecture we would only do this if $f_{x_1 x_2 \cdots x_k} = 1$ or $f_{x_1 x_2 \cdots x_k} = 0$.

Since there is no limit to the number of inputs that the precomputation logic is a function of, we need to monitor its size in order to ensure it does not get very large. In the sequel we describe a branching algorithm that selects the "best" subset of inputs. The pseudo-code is shown in Figures 7.17 and 7.18.

The procedure **SELECT_LOGIC** receives as arguments the function

```
1.      SELECT_LOGIC( f, k ):
2.      {
3.          BEST_IN_COST = 0 ;
4.          SELECTED_SET = ϕ ;
5.          SELECT_RECUR( f, ϕ, X, k ) ;
6.          return( SELECTED_SET ) ;
7.      }
```

Figure 7.17: Input selection for the complete input disabling architecture.

f and the desired number of inputs k to select. **SELECT_LOGIC** calls the recursive procedure **SELECT_RECUR** with four arguments. The first is the function to precompute. The second argument D corresponds to the set of input variables currently selected. The third argument Q corresponds to the set of "active" variables, which may be selected or discarded. Finally, the argument k corresponds to the number of variables we want to select.

If $|D| + |Q| < k$ it means that we have dropped too many variables in the earlier levels of recursion and we will not be able to select a subset of k input variables.

When k inputs have been selected, we compute the cofactors of f with respect to all combinations over the input variables currently in D. We want to keep those cofactors that have a high probability of being 0 or 1. The cost function is the fraction of exact cofactors found (exact meaning that the selected inputs determine the value of f) plus a factor $\frac{\text{size}(g_e)}{\text{size}(g_a)}$ times the fraction of approximate cofactors found (with these cofactors we still need variables *not* in D to be able to precompute f). The factor $\frac{\text{size}(g_e)}{\text{size}(g_a)}$ tries to measure how much more complex the precomputation logic will be by selecting these approximate factors.

We can tune the value of α thus controlling how many approximate cofactors we select. The more we select, the more input combinations will be in the precomputation logic therefore increasing the fraction of the time that we will be disabling the input registers. On the other hand, the logic will be more complex since we will need more input variables. Note that in the extreme case of $\alpha = 0$, the input selection will be the same as in subset input disabling architecture as all the selected input combinations depend only on the inputs that are in subset D.

7. PRECOMPUTATION

```
1.      SELECT_RECUR( f, D, Q, k ):
2.      {
3.          if( |D| + |Q| < k )
4.              return ;
5.          if( |D| == k ) {
6.              exact = approx = 0;
7.              g_e = g_a = 0;
8.              foreach combination c over all variables in D {
9.                  if(prob(f_c) == 1 or prob(f_c) == 0) {
10.                     exact = exact + 1;
11.                     g_e = g_e + c;
12.                     g_a = g_a + c;
13.                     continue ;
14.                 }
15.                 if(prob(f_c) > 1 − α) {
16.                     approx = approx + 1;
17.                     g_a = g_a + c·f_c;
18.                 }
19.                 if(prob(f_c) < α) {
20.                     approx = approx + 1;
21.                     g_a = g_a + c·f̄_c;
22.                 }
23.             }
```

24. $cost = (exact + \frac{\text{size}(g_e)}{\text{size}(g_a)} \times approx)/2^{|D|}$;

```
25.         if( cost > BEST_IN_COST) {
26.             BEST_IN_COST = cost ;
27.             SELECTED_SET = D ;
28.         }
39.         return ;
30.     }
31.     select next x_i ∈ Q ;
32.     SELECT_RECUR( f, D ∪ x_i, Q − x_i, k ) ;
33.     SELECT_RECUR( f, D, Q − x_i, k ) ;
34. }
```

Figure 7.18: Recursive procedure for input selection for the complete input disabling architecture.

We store the selected set corresponding to the maximum value of the cost function.

Selecting a Subset of Inputs: Approximate Method

The previous method is very expensive as it is exponential in the number of primary inputs. The approximate method we propose to select the "best" subset of inputs is the same as for the subset inputs disabling architecture. For every primary input x_i, we compute:

$$p_i = prob(U_{x_i}f) + prob(U_{x_i}\overline{f}) \qquad (7.13)$$

and select the k inputs corresponding to smaller values of p_i.

The difference now is that given this subset of inputs D, we compute the cofactors of f with respect to every combination c in D. If $prob(f_c) > 1 - \alpha$ we include $c \cdot f_c$ in g_1. If $prob(f_c) < \alpha$ we include $c \cdot \overline{f_c}$ in g_2.

Implementing the Logic

The Boolean operations of OR and cofactoring required in the input selection procedure can be carried out efficiently using ROBDDs. In the pseudo-code of Figure 7.18 we show how to obtain the $g_1 + g_2$ function. We also need to compute g_1 and g_2 independently. We do this in exactly the same way, by including in g_1 the cofactors corresponding to probabilities close to 1 and in g_2 the cofactors corresponding to probabilities close to 0.

Again, given ROBDDs for g_1 and g_2, these can be converted into a multiplexor-based network or into a sum-of-products cover.

7.2.4 Simplifying the Original Combinational Logic Block

Whenever g_1 or g_2 evaluate to a 1, we will not be using the result produced by the original combinational logic block A, since the value of f will be set by either g_1 or g_2. Therefore all input combinations in the precomputation logic are new don't-care conditions for this circuit and we can use this information to simplify the logic in block A, thus leading to a reduction in area and consequently to a further reduction in power dissipation.

7. PRECOMPUTATION 139

CIRCUIT NAME	ORIGINAL				
	I	O	LTS	D	P
9sym	9	1	303	19.6	1828
Z5xp1	7	10	163	34.8	1533
alu2	10	6	501	42.2	2988
apex2	39	3	330	15.6	1978
cm138	6	8	34	5.8	232
cm152	11	1	30	6.4	427
cm162	14	5	66	9.8	540
cmb	16	4	75	7.0	653
dalu	75	16	1271	46.0	7003
mux	21	1	65	9.8	806
sao2	10	4	181	24.6	1001

Table 7.3: Power reductions in sequential precomputation using the complete input disabling architecture.

7.2.5 Multiple-Output Functions

The extension of the previous algorithms for multiple-output functions is done in exactly the same way as for the subset input disabling architecture. We use the exact method of Figure 7.6 and the approximate method of Figure 7.8. When precomputing a subset of outputs, the problem of logic duplication of Section 7.1.4 remains the same for this architecture.

7.2.6 Experimental Results for the Complete Input Disabling Architecture

We present in Tables 7.3 and 7.4 the power saving results using sequential precomputation under the complete input disabling architecture. Again we are using circuits taken from the MCNC benchmark set and have assumed that the inputs to the circuits are outputs of flip-flops.

In Table 7.3, under ORIGINAL, we present for each circuit the number of inputs (I), outputs (O), literals (LTS), the maximum delay in nanoseconds (D), and power (P) of the original circuit. In Table 7.4, we present results obtained with the complete input disabling architecture. Under PRECOMPUTE LOGIC we give the number of inputs in the selected set (I), number of precomputed outputs (O), literals (LTS) and delay (D) of the precomputation logic. Under OPTIMIZED, we give the delay (D) and power (P) of the optimized precomputed network, and the percent reduction (RED) in power. All power estimates are

Circuit	Precompute Logic				Optimized		
Name	I	O	LTS	D	D	P	RED
9sym	7	1	53	13.8	20.4	1255	31.3
Z5xp1	2	1	3	2.8	34.8	1325	13.6
alu2	5	3	24	8.6	44.0	2648	11.4
apex2	10	3	23	7.2	27.2	984	50.0
cm138	3	8	4	5.4	7.4	136	41.4
cm152	9	1	26	7.8	9.2	301	29.5
cm162	9	5	24	4.8	10.8	370	31.5
cmb	8	4	40	5.4	8.8	224	65.7
dalu	6	16	68	11.6	46.3	3720	46.9
mux	1	1	1	1.6	11.2	539	33.1
sao2	2	4	5	2.4	23.6	406	59.3

Table 7.4: Power reductions in sequential precomputation using the complete input disabling architecture (contd).

in micro-Watt and are computed using the techniques described in Chapter 4. A zero delay model, a clock frequency of 20MHz and a supply voltage of 5V was assumed. The rugged script of SIS [7] was used to optimize the precomputation logic.

Note that the delay of the precomputation logic is added to the delay of the *previous* stage in sequential precomputation. The delay numbers in the third to last column correspond to the critical delay of the optimized circuit which includes the output AND-OR gate (cf. Figure 7.14). However, the use of don't-care conditions to optimize the circuit once the precomputation logic has been determined can reduce the delay of the optimized circuit.

In Table 7.5 we compare the complete and subset input disabling precomputation architectures. The best results obtained by both methods for each of the examples is given. The precomputation logic in the complete input disabling method is typically larger than in the subset input disabling method, however the first can achieve larger power reductions. The reason for this is twofold. First the probability of the precomputation logic can be higher for the complete input disabling architecture. Secondly, the original circuit is simplified due to the don't-care conditions in the complete input disabling architecture.

7. PRECOMPUTATION

Circuit Name	Orig. Power	Subset Input Disable				Complete Input Disable			
		LTS	D	P	% RED	LTS	D	P	% RED
9sym	1828	40	11.0	1610	11.9	53	13.8	1255	31.3
Z5xp1	1533	3	2.8	1390	9.3	3	2.8	1325	13.6
alu2	2988	8	4.0	2683	10.2	24	8.6	2648	11.4
apex2	1978	15	5.3	1196	39.5	23	7.2	984	50.0
cm138	232	3	2.6	146	37.0	4	5.4	136	41.4
cm152	427	5	2.6	395	7.5	26	7.8	301	29.5
cm162	540	2	1.4	466	13.7	24	4.8	370	31.5
cmb	653	13	3.8	436	33.2	40	5.4	224	65.7
cordic	928	13	5.2	798	14.0	114	12.2	553	40.0
dalu	7003	16	5.6	4292	38.7	68	11.6	3720	46.9
mux	806	0	0	591	26.7	1	1.6	539	33.1
sao2	1001	2	1.4	446	55.4	5	2.0	406	59.3

Table 7.5: Comparison of power reductions between complete and subset input disabling architectures.

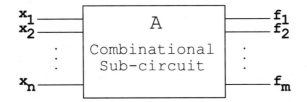

Figure 7.19: Original combinational sub-circuit.

7.3 Combinational Precomputation

The architectures described so far apply only to sequential circuits. We now describe precomputation for combinational circuits.

7.3.1 Combinational Logic Precomputation

Given a combinational circuit, any sub-circuit within the original circuit can be selected to be precomputed. Assume that we select a sub-circuit with n inputs and m outputs as shown in Figure 7.19. In an effort to reduce switching activity, the algorithm will "turn off" a subset of the n inputs using the circuit shown in Figure 7.20. The figure shows p inputs being "turned off", where $1 \leq p < n$.

The term "turn off" means different things according to the type of

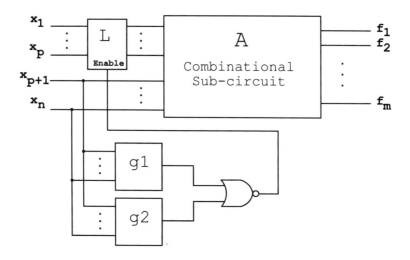

Figure 7.20: Sub-circuit with input disabling circuit.

circuit style that is being used. If the circuit is built using static logic gates, then "turn off" means prevent changes at the inputs from propagating through block L to the sub-circuit (block A) thus reducing the switching activity of the sub-circuit. In this case block L may be implemented using one of the transparent latches shown in Figure 3.7. If the circuit is built using dynamic logic, then "turn off" means prevent the outputs of block L from evaluating high no matter the value of the inputs. This can be implemented simply by using 2-input AND gates where one of the inputs is the enable signal.

Blocks g_1 and g_2 determine when it is appropriate to turn off the selected inputs. The selected inputs may be "turned off" if the static value of all the outputs, f_1 through f_m, are independent of the selected inputs. To fulfill this requirement, outputs g_1 and g_2 are required to satisfy Equations 7.1 and 7.2. If either g_1 or g_2 is high, the inputs may be "turned off". If they are both low, then the selected inputs are needed to determine the outputs, and the circuit is allowed to work normally.

There are two interesting cases of combinational precomputation that have differing merits and demerits. We discuss these cases in the next two sections.

7. PRECOMPUTATION

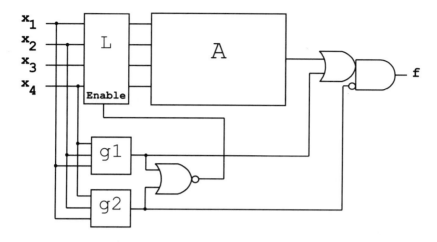

Figure 7.21: Complete input disabling for combinational circuits.

7.3.2 Precomputation at the Inputs

The sub-circuit considered in Figure 7.19 can be precomputed as shown in Figure 7.20. The algorithms presented in Section 7.1 for the subset input disabling architecture are directly applicable in this case. A subset of inputs $x_1, \cdots, x_p$ can be selected that achieves maximal power savings.

In order to ensure power savings the $x_1, \cdots, x_p$ inputs should be delayed such that new values arrive at the transparent latches *after* the new value of the enable signal arrives. Else, these new values may propagate through to block A causing unnecessary transitions.

The complete input disabling architecture can also be used for combinational circuits. This is illustrated in Figure 7.21. The algorithms described in Section 7.2 can be applied directly to synthesize the precomputation logic.

7.3.3 Precomputation for Arbitrary Sub-Circuits in a Circuit

In the general case we wish to synthesize precomputation logic for arbitrary sub-circuits as illustrated in Figure 7.22.

In this case algorithms are needed to accomplish several tasks. First, an algorithm must divide the circuit into sub-circuits. Then for each sub-circuit, algorithms must: a) select the subset of inputs to "turn off," and b) given these inputs, produce the logic for g in Figure 7.22, where $g = g_1 + g_2$.

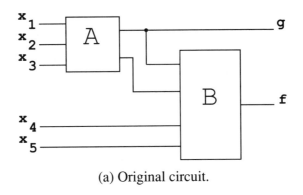

(a) Original circuit.

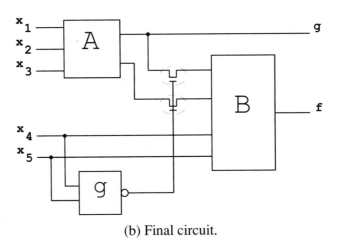

(b) Final circuit.

Figure 7.22: Combinational logic precomputation.

For each of these steps, the goal is to maximize the savings function

$$net\ savings = \sum_{\text{all subcircuits}} (savings(A) - cost(L) - cost(g)) \qquad (7.14)$$

We have to partition the original circuit into sub-circuits so that Equation 7.14 is maximized. The original circuit can be divided into a set of maximum-sized, single-output sub-circuits. A maximum-sized, single-output sub-circuit is a single-output sub-circuit such that no set of nodes from the original circuit can be added to this sub-circuit without creating a multiple-output sub-circuit. An equivalent way of saying this is, the circuit can be divided into a minimum number of single-output sub-circuits. Such a set exists

7. PRECOMPUTATION

```
1.      GET_SINGLE_OUTPUT_SUBCIRCUITS( circuit ):
2.      {
3.          arrange nodes of circuit in depth-first order outputs to inputs;
4.          foreach node in depth order ( node ) {
5.              if ( node is a primary output ) {
6.                  subcircuit = create_new_subcircuit();
7.                  mark node as part of subcircuit;
8.              }
9.              else {
10.                 check every fanout of node;
11.                 if ( all fanouts are part of the same subcircuit )
12.                     subcircuit = subcircuit of the fanouts;
13.                 else
14.                     subcircuit = create_new_subcircuit();
15.                 mark node as part of subcircuit;
16.             }
17.         }
18.     }
```

Figure 7.23: Procedure to find the minimum set of single-output subcircuits.

and is unique for any legal circuit. A linear-time algorithm for determining this set is given in Figure 7.23.

Next, note that there is no need to analyze any sub-circuit that is composed of only a part of one of these maximum-sized, single-output sub-circuits. If a part of a single-output sub-circuit including the output node is in some sub-circuit to be analyzed, then the rest of the nodes of the single-output sub-circuit can be added to the sub-circuit *at no cost* since the outputs remain the same. Adding these nodes can only result in more savings. Further, if a part of a single-output sub-circuit not including the output node is in some sub-circuit to be analyzed, then the rest of the nodes of the single-output sub-circuit can be added to the sub-circuit because the precomputability of the outputs can only become less restrictive. Therefore, even in the worst case, the disable logic can be left the same so that there is no additional cost yet additional savings are achieved because of the additional nodes.

Based upon this theory, an algorithm to synthesize precomputation logic would 1) create the set of maximum-sized, single-output sub-circuits, 2) try different combinations of these sub-circuits, and 3) determine the combinations that yield the best net savings. Given the maximum-sized single-output sub-circuits, we use the algorithms of Sections 7.1 or 7.2 to determine a subset of the sub-circuits and a selection of inputs to each sub-circuit that results in relatively simple precomputation logic and maximal power savings.

Note that in this strategy the waveforms that appear at the inputs to a latch can be arbitrary. The arrival time at the input should be *later* than the arrival time of the enable signal so that unnecessary transitions are not propagated through the latch. In the example shown in Figure 7.22, the worst-case delay of the g block plus the arrival time of inputs x_4 or x_5 should be less than the best-case delay of logic block A plus the arrival time of the inputs x_1, x_2, or x_3. The *arrival time* of an input is defined as the time at which the input settles to its steady state value [5, p. 229]. If the delay constraint is not met, then it may be necessary to delay the x_1, x_2, and x_3 inputs with respect to the x_4 and x_5 inputs in order to get the switching activity reduction in logic block B.

7.3.4 Experimental Results for the Combinational Precomputation Architecture

In Table 7.6 we present results on combinational precomputation. The symbolic simulation method of Chapter 3 was used to obtain the power estimates of the combinational circuits with transparent latches. Again, a zero delay model, a clock frequency of 20MHz and a supply voltage of 5V was assumed.

The same circuits as for the complete input disabling architecture were selected to provide a comparison between the combinational and sequential architectures. The number of inputs (I), outputs (O), literal count (LTS) and delay (D) of the precomputation logic are given under PRECOMPUTE LOGIC. The critical delay of the final precomputed network which includes additional delay introduced due to the transparent latches, the precomputation logic, and any delaying of inputs is given in the third to last column. As can be observed from Table 7.6, substantial reductions in power can be obtained with small increases in delay.

Circuit	Original		Precompute Logic				Optimized		
Name	D	P	I	O	LTS	D	D	P	% Red
9sym	19.6	1625	7	1	53	11.0	32.4	1960	-21.1
Z5xp1	34.8	1375	2	7	3	2.8	36.2	1339	2.6
alu2	42.2	2763	8	6	30	6.8	50.2	2792	-1.0
apex2	15.6	1094	4	3	5	2.6	28.2	948	13.3
cm138	5.8	97	3	8	4	2.6	8.8	68	29.9
cm152	6.4	179	5	1	4	2.6	11.4	183	-2.2
cm162	9.8	225	1	4	0	0.0	13.6	177	21.3
cmb	7.0	293	5	5	14	2.8	13.6	194	33.8
dalu	46.0	5312	5	16	18	5.4	61.6	4050	23.8
mux	9.8	334	1	1	0	0.0	15.0	168	49.7
sao2	24.6	776	2	4	2	1.4	29.2	545	29.8

Table 7.6: Power reductions using combinational precomputation.

7.4 Multiplexor-Based Precomputation

In this section, we describe an additional precomputation architecture. This *Multiplexor-Based* precomputation architecture. is applicable to *all* logic circuits and does not require, for instance, that the inputs should be in the observability don't-care set in order to be disabled, which was the case for all the previous architectures.

All logic functions can be written in terms of its Shannon expansion. For a function f with inputs $X = \{x_1, \cdots, x_n\}$, we can write:

$$f = x_1 \cdot f_{x_1} + \overline{x_1} \cdot f_{\overline{x_1}} \qquad (7.15)$$

where f_{x_1} and $f_{\overline{x_1}}$ are the cofactors of f with respect to x_1 and $\overline{x_1}$.

Figure 7.24 shows an architecture based on Equation 7.15. We implement the functions f_{x_1} and $f_{\overline{x_1}}$. Depending on the value of x_1, only one of the cofactors is computed while the other is disabled by setting the load-enable signal of its input register. The input x_1 drives the select line of a multiplexor which chooses the correct cofactor.

The main advantage of this architecture is that it applies to *all* logic functions. The input x_1 in the example was chosen for the purpose of illustration. In fact, any input $x_1, \cdots, x_n$ could have been selected. Unlike the architectures described earlier, we do not require that the inputs being disabled should be don't-cares for the input conditions which we are precomputing. In other words, the inputs being disabled do not have to be in the observability

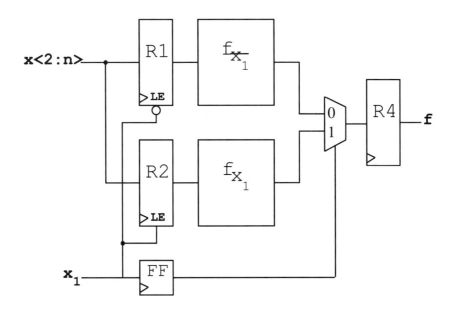

Figure 7.24: Precomputation using the Shannon expansion.

don't-care set. A disadvantage of this architecture is that we need to duplicate the registers for the inputs not being used to turn off part of the logic. On the other hand, no precomputation logic functions have been added to the circuit.

This precomputation architecture was the subject of one of the problem sets in the graduate course *Computer-Aided Design of Integrated Circuits* at MIT (course 6.373) in the Spring term of 1995. Students were asked to develop an algorithm to select the best primary input to use for the architecture of Figure 7.24. For half of the eight two-level benchmark circuits used, power savings of more than 40% were achieved.

7.5 Conclusions

We have presented synthesis algorithms that can be used to optimize a given combinational or sequential logic circuit for low power dissipation by adding "precomputation logic" which reduces unnecessary transitions in large parts of the given circuit. The output response of a sequential circuit is precomputed one clock cycle before the output is required, and this knowledge is exploited to reduce power dissipation in the succeeding clock cycle. As opposed to power-down techniques applied at the system level, transition reduction is achieved on a per clock cycle basis.

Several different architectures that utilize precomputation logic were presented. Precomputation increases circuit area and can adversely impact circuit performance. In order to keep area and delay increases small, it is best to synthesize precomputation logic which depends on a small set of inputs.

Precomputation works best when there are a small number of complex functions corresponding to the logic block A of Figures 7.2 and 7.14. If the logic block has a large number of outputs, then it may be worthwhile to selectively apply precomputation-based power optimization to a small number of complex outputs. This selective partitioning will entail a duplication of combinational logic and registers, and the savings in power is offset by this duplication.

Other precomputation architectures are being explored, including the architectures of Section 7.4, and those that rely on a history of previous input vectors. More work is required in the automation of a logic design methodology that exploits multiplexor-based, combinational and multiple-cycle precomputation.

In the next chapter we describe techniques to explore data-dependent power-down at the register-transfer and behavioral levels.

References

[1] M. Alidina, J. Monteiro, S. Devadas, A. Ghosh, and M. Papaefthymiou. Precomputation-Based Sequential Logic Optimization for Low Power. *IEEE Transactions on VLSI Systems*, 2(4):426–436, December 1994.

[2] P. Ashar, S. Devadas, and K. Keutzer. Path-Delay-Fault Testability Properties of Multiplexor-Based Networks. *INTEGRATION, the VLSI Journal*, 15(1):1–23, July 1993.

[3] R. Brayton, R. Rudell, A. Sangiovanni-Vincentelli, and A. Wang. MIS: A Multiple-Level Logic Optimization System. *IEEE Transactions on Computer-Aided Design*, 6(6):1062–1081, November 1987.

[4] R. Bryant. Graph-Based Algorithms for Boolean Function Manipulation. *IEEE Transactions on Computers*, C-35(8):677–691, August 1986.

[5] S. Devadas, A. Ghosh, and K. Keutzer. *Logic Synthesis*. McGraw Hill, New York, NY, 1994.

[6] J. Monteiro, J. Rinderknecht, S. Devadas, and A. Ghosh. Optimization of Combinational and Sequential Logic Circuits for Low Power Using Precomputation. In *Proceedings of the 1995 Chapel Hill Conference on Advanced Research on VLSI*, pages 430–444, March 1995.

[7] E. Sentovich, K. Singh, C. Moon, H. Savoj, R. Brayton, and A. Sangiovanni-Vincentelli. Sequential Circuit Design Using Synthesis and Optimization. In *Proceedings of the International Conference on Computer Design: VLSI in Computers and Processors*, pages 328–333, October 1992.

[8] A. Shen, S. Devadas, A. Ghosh, and K. Keutzer. On Average Power Dissipation and Random Pattern Testability of Combinational Logic Circuits. In *Proceedings of the International Conference on Computer-Aided Design*, pages 402–407, November 1992.

[9] V. Tiwari, P. Ashar, and S. Malik. Guarded Evaluation: Pushing Power Management to Logic Synthesis/Design. In *Proceedings of the International Symposium on Low Power Design*, pages 221–226, April 1995.

Chapter 8

High-Level Power Estimation and Optimization

The methods of Chapter 7 are limited by the predefined logical structure of the circuit. Techniques at a higher abstraction levels, namely behavioral and register-transfer levels, yield a potentially larger optimization impact since these constraints do not yet exist. Further, logic level descriptions are too detailed to allow optimization methods to be applied to large designs.

Behavioral synthesis comprises of the sequence of steps by means of which an algorithmic specification is translated into hardware. These steps involve breaking down the algorithm into primitive operations, and associating each operation with the time interval in which it will be executed (called operation scheduling) and the hardware functional module that will execute it (called hardware assignment). Clock period constraints, throughput constraints and hardware resource constraints make this a non-trivial optimization problem [5].

At the register-transfer level (RTL), the circuit is described in terms of functional modules, of different levels of complexity, and how they are interconnected. In general, there are two main circuit components. The first is the datapath, which is composed of the functional modules that actually do the computation, registers (or memory) to hold the computation results, and multiplexors that direct the flow of data in the circuit. The second is the controller, which generates the sequence of control signals for the datapath, indicating which operation each functional unit should execute, controlling the load and output enable of the registers, selecting memory addresses and multiplexor inputs.

In this chapter we describe most of the significant optimization techniques for low power that have been proposed at the behavioral level. Techniques to estimate the power consumption of the RTL circuit are required so that different designs can be compared. We begin by presenting a survey of power estimation techniques at the RT level.

8.1 Register Transfer Level Power Estimation

In this section we focus on power estimation techniques at the RT level. Estimation techniques at higher levels, i.e., behavioral, do exist [16, 12, 1]. However, the accuracy of these tools is necessarily very limited given the reduced information about circuit implementation available at this level.

The general approach to compute the power dissipation of a logic circuit described at the RT level is to estimate the power of each of the modules that make up the circuit. Even if the details of the specific implementation of modules is known, for the sake of efficiency a much simpler model for the module is used. These models are additional data stored in the library of modules. Since we are using less detailed information about the circuit we cannot obtain power estimates as accurate as at the logic level. The designer will be willing to lose some accuracy in order to be able to handle larger designs, provided the relative power estimates between two designs are accurate enough to determine which one is more power dissipative.

We present techniques that have been proposed to model the main components of the circuit at the RT level.

8.1.1 Functional Modules

The basic model used by most power estimation techniques for a functional module is of the form [17]:

$$P = C_{eff} K V_{DD}^2 f$$

where V_{DD} is the voltage of the power supply for the module and f is the clock frequency at which it operates.

K is the scaling factor, as a general assumption is that power scales with area. For instance, if N is the word length, $K = N$ for an adder and $K = N^2$ for a multiplier.

8. HIGH-LEVEL POWER ESTIMATION AND OPTIMIZATION

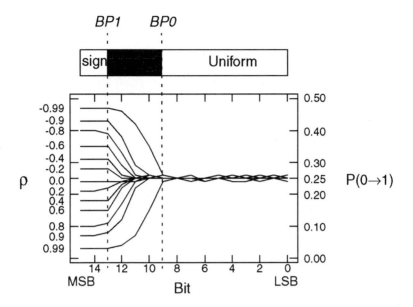

Figure 8.1: Bit transition probability as a function of temporal correlation.

Lastly, the factor C_{eff} is the average switched capacitance per bit slice of the functional unit. C_{eff} is computed beforehand for each type of functional module and stored in the module library. In general, C_{eff} is obtained by simulating each different module using a large number of input vectors, randomly generated using a uniform distribution.

This power estimation model for functional modules is extended in [8] in two ways. First, the authors note that it may require more than a single parameter to describe how the switched capacitance varies for different sizes of a given type of module. For example, a register file with R registers and word length N should be described by capacitance values that scale with R to model the control logic, a function of how many words can be stored, N for the input/output buffers, and RN, the number of storage cells. The switched capacitance is then given by

$$C_0 R + C_1 N + C_2 NR = \boldsymbol{C}_{eff} \cdot \boldsymbol{K}$$

where $\boldsymbol{C}_{eff} = [C_0 \ C_1 \ C_2]$ and $\boldsymbol{K} = [R \ N \ RN]^T$. Again, the values C_0, C_1 and C_2 are computed beforehand by simulation.

A second observation made in [8] is that assuming a uniform distribution for all the inputs to the module can be very inaccurate. Figure 8.1 shows how the switching probability as a function of the temporal correlation ρ of consecutive data values varies from the most to the least significant bit in the datapath, assuming a two's-complement representation. As can be observed, independently of the correlation value, the least significant bits present a uniform probability of switching. On the other hand, the switching probability of the most significant bits ($> BP1$) is highly dependent on the correlation value. These bits can have very low or very high switching probabilities as ρ varies from 1 to -1.

In [8] it is proposed that different capacitive coefficients be used for these two distinct regions. For the least significant bits ($< BP0$), the uniform approach is used. For the most significant bits ($> BP1$), several capacitive coefficients are needed, one for each possible signal transition. The total switched capacitance is then given by

$$C_T = \frac{N_U}{N}[C_U \cdot K] + \frac{N_S}{N}\left[\sum_{S\in(++,+-,-+,--)} P(S)\, C_S \cdot K\right] \quad (8.1)$$

where $P(S)$ is the probability of having a change S in signal, and $\frac{N_U}{N}$ and $\frac{N_S}{N}$ are the fraction of bits in the uniform and signal regions respectively, obtained from $BP0$ and $BP1$. These three values are obtained by simulation.

The previous approaches can model the glitching activity that is generated within a module by taking it into account during the characterization phase of the module. However, it assumes that the inputs to the modules are glitch free. A different power model is proposed in [19] that is able to handle glitching activity at the inputs.

In this approach the glitching activity at the output of a module with two inputs A and B is computed using

$$Gl_{out} = f_{gl_1}(M_A, M_B) \cdot f_{gl_2}(SD_A, SD_B) \cdot f_{gl_3}(TC_A, TC_B) \cdot f_{gl_4}(SC_{A,B}) \cdot f_{gl_5}(Gl_A, Gl_B) \quad (8.2)$$

where M_A, SD_A and TC_A represent respectively the mean, standard deviation and temporal correlation of the word-level value of input A, and Gl_A is the glitching present at input A. Similarly for input B. $SC_{A,B}$ represents the spatial correlation between inputs A and B. These statistical values are computed from cycle-based simulation of the RTL circuit. f_{gl_1} through f_{gl_5} are piecewise linear functions constructed to fit experimental points obtained from simulation during the characterization process. The power dissipation of the module is computed using an expression similar to Equation 8.2.

8. HIGH-LEVEL POWER ESTIMATION AND OPTIMIZATION 155

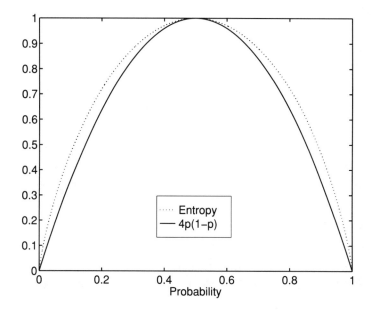

Figure 8.2: Entropy and switching probability of a Boolean signal as a function of its static probability.

The methods described so far are all simulation-based. Two probabilistic techniques are described in [15] and [11]. Both these techniques are based on entropy and their focus is to derive implementation-independent measures of the signal activity in the circuit.

The first assumption both methods make is that the average power dissipation of the module can be approximated by

$$P_{avg} \propto \sum_{\substack{\text{all gates } g \\ \text{in module}}} C_g N_g \approx \mathcal{C} \times \mathcal{N} \quad (8.3)$$

where $\mathcal{C}$ represents the total capacitance and $\mathcal{N}$ the average node switching activity.

The entropy of a discrete variable x that can assume n different values, each having a probability p_i of occurring, is given by

$$H(x) = \sum_{i=1}^{n} p_i \log_2 \frac{1}{p_i}. \quad (8.4)$$

8.1. REGISTER TRANSFER LEVEL POWER ESTIMATION

The entropy of a Boolean signal x having a probability p_1 of being 1 is

$$H(x) = p_1 \log_2 \frac{1}{p_1} + (1 - p_1) \log_2 \frac{1}{1 - p_1}. \tag{8.5}$$

The plot of the entropy of a Boolean signal as a function of the static probability of the signal being 1 is shown in Figure 8.2.

Ignoring temporal correlation, the switching probability of a signal is given by $2p_1(1-p_1)$. The plot of twice this amount is also shown in Figure 8.2 and as it can be seen it is very close to the entropy. Techniques [15] and [11] use the average entropy $\mathcal{H}$ as the measure for the switching activity,

$$P_{avg} \propto \mathcal{C} \times \mathcal{H}. \tag{8.6}$$

The two techniques differ in the way they estimate $\mathcal{C}$ and $\mathcal{H}$.

In [15] it is shown empirically that the entropy decreases quadratically with the module depth. This observation is used to arrive at the expression

$$\mathcal{H} \approx \frac{\frac{2}{3}}{n+m}(H_i + 2H_o) \tag{8.7}$$

where H_i and H_o are the sum of the entropies of the input and output signals respectively, n is the number of inputs and m the number of outputs. H_i can be computed directly from the input signal probabilities. However, the entropy of output signals H_o is computed from signal statistics obtained by simulation.

The total capacitance is represented in [15] by an estimate of the module area $\mathcal{A}$, $P_{avg} \propto \mathcal{A} \times \mathcal{H}$. The authors propose an upper and lower bound for $\mathcal{A}$, which are functions of the output entropy and empirically found functions of the number of inputs n,

$$0.4 H_o \left(\frac{n}{\log_{10} n} \right) \leq \mathcal{A} \leq 2 H_o (n \log_{10} n). \tag{8.8}$$

A different expression for calculating $\mathcal{H}$ is given in [11],

$$\mathcal{H} \approx \frac{H_i - H_o}{\ln \frac{H_i}{H_o}}. \tag{8.9}$$

The entropy at the output H_o is computed from the entropy at the input, the number of levels in the circuit N and an *information scaling factor* f_{eff},

$$H_o = \frac{H_i}{f_{eff}^{N/2}}. \tag{8.10}$$

8. HIGH-LEVEL POWER ESTIMATION AND OPTIMIZATION

The value of f_{eff} is dependent on the functionality of the gates in the module. In case the structure of the module is known, more specific expressions for $\mathcal{H}$ can be used and are provided in [11].

It is assumed in [11] that $\mathcal{C}$ is available from the module library.

The methods of [15] and [11] can be very efficient, but given all the required approximations and the fact that the methods ignore issues such as glitching implies that these methods are not very robust in terms of accuracy.

8.1.2 Controller

Estimating the power dissipation of the controller at the RT level is significantly more difficult than for the datapath modules because the controller does not have the same regularity, and further, its implementation is very application dependent. At this level there is no information about its final implementation, the controller is described in terms of a State Transition Graph (STG). Therefore, the information available for the power estimation process is the number of primary inputs N_{PI}, primary outputs N_{PO} and states N_{States} in the STG and the structure of the STG.

The controller is generally implemented as a Finite State Machine (FSM) (cf. Figure 4.3). However, there are several different ways the combinational logic block can be implemented.

The power estimation technique described in [9] has capacitance models for three different types of the FSM implementation: Read-Only Memory (ROM)-based, Programmable Logic Array (PLA)-based and random logic. The user has to specify which case applies for his/her design.

For ROM-based FSM implementation, the model proposed in [9] for the average switched capacitance is

$$C_T = C_0 + C_1 N_I 2^{N_I} + C_2 P_O N_O 2^{N_I} + C_3 P_O N_O + C_4 N_O \qquad (8.11)$$

with $N_I = N_{PI} + N_S$ and $N_O = N_{PO} + N_S$, where N_S is the number of state lines. P_O is the probability of the output signals being 1. In practice, P_O is assumed to be 1/2. Again, the capacitive coefficients C_0 through C_5 are obtained by a characterization process where FSMs of different complexity are used. In Equation 8.11, $C_1 N_I 2^{N_I}$ represents the capacitance of the input plane, $C_2 P_O N_O 2^{N_I}$ the output plane and both $C_3 P_O N_O$ and $C_4 N_O$ are related to the output buffering circuitry.

A similar expression is derived for a PLA-based FSM implementation. For FSMs that use random logic for the combinational logic block, the switched capacitance is more difficult to predict because these are much less regular than ROM or PLA implementations. The expression proposed in [9] is

$$C_T = C_0 \alpha_I N_I N_M + C_1 \alpha_O N_O N_M \tag{8.12}$$

where N_M is the number of minterms in the combinational logic block and α_I and α_O are the input and output switching activities respectively. N_M is estimated by assuming a random encoding of the states and performing a quick logic minimization of the combinational logic. The input and output switching activities (α_I and α_O) are obtained from the functional simulation of the circuit.

The estimation method proposed in [19] is especially concerned with determining the amount of glitching present in the primary outputs of the controller. These signals control datapath modules and can lead to spurious transitions in lines with high capacitance, such as data buses. In [19], the control signals are assumed to be in the form

$$contr = \sum_i x_i \cdot \prod_j C_{ij} \tag{8.13}$$

where x_i represents the state the controller is in, and C_{ij} the primary inputs to the controller.

A cycle-based simulation of the controller is done giving both the zero-delay switching activity and signal statistics. From these signal statistics, glitching at the control signals is estimated, both from the glitches present in each of the minterms in Equation 8.13 and their disjunction. Glitching depends heavily on timing information, which is not available at this level. The technique of [19] divides the signals into early arriving signals (such as state lines) and late arriving signals (such as comparator outputs) to have some information about delays. When a signal is neither early or late arriving, the most pessimistic situation is used.

The final power estimation is obtained assuming a straightforward implementation of the expressions in Equation 8.13. The zero-delay switching activity plus the estimated amount of glitching is multiplied by typical capacitance values for inverters and 2-input AND and OR gates.

8.1.3 Interconnect

At the RT level the effects of the interconnect on power cannot be neglected as many of the signal lines may span a significant portion of the final circuit. The signal activity can be estimated from functional simulation. Additionally, the capacitance of the lines has to be computed. However, in this early stage of design, no more than a crude estimate can be obtained.

Methods to predict the amount of interconnect area from the amount of active area have been proposed in [3, 4]. From the interconnect area, interconnect capacitance can be computed for a particular fabrication process. In [10, 23] the amount of active area for each module is assumed to be available from the module library and therefore the computation of an estimate for the interconnect capacitance is immediate.

In [1], a model for the switched capacitance in the interconnect lines is built from the statistics obtained over a large number of design examples.

8.2 Behavioral Level Synthesis for Low Power

Decisions taken during behavioral synthesis have a far reaching impact on the power dissipation of the resulting hardware. We describe behavioral synthesis techniques that have been proposed aiming for a power optimized RT level circuit.

Ideally, we start with an algorithmic description of the design in some hardware description language. To exemplify, consider the simple circuit dealer described in Silage [7] in Figure 8.3. This figure represents one cycle of the dealer, where @1 refers to the value of the variable in the previous cycle and @@1 are initialization values.

The first step is in general translating this textual description into a Control Data Flow Graph (CDFG). The CDFG corresponding to the dealer circuit of Figure 8.3 is shown in Figure 8.4. The nodes in the CDFG are operations in the circuit and the edges indicate precedence conditions among the operations. Most high-level synthesis and optimization techniques operate on the CDFG that describes the circuit.

8.2.1 Transformation Techniques

Even before the start of the synthesis process, transformations of the circuit have been proposed to reduce power dissipation. Among the most effective transformation techniques are those that allow for the reduction of the power supply voltage, given its quadratic relationship to power. One approach is to improve throughput by exploiting concurrency via pipelining.

```
func main(PresentSuit, NoSuit, Incr, Limit, DeckSize: int<8>)
Card, Avalue: int<8> =
    begin
        Card@@1 = 0;
        Avalue@@1 = 0;
        Card = if (PresentSuit == NoSuit) ->
            if (Limit < Card@1) ->
                Card@1 - Limit
            ||
                Card@1 + Incr
            fi
        ||
            if (Card@1 + Incr >= DeckSize) ->
                1
            ||
                Card@1 + Incr
            fi
        fi;
        Avalue = 1 + Avalue@1;
    end;
```

Figure 8.3: Silage description of the `dealer` **circuit.**

This enables the hardware to be operated at lower clock frequencies, and thereby at lower voltages [1, 6]. The disadvantage of using pipelining is that the latency of the circuit increases and we may require aditional hardware due to having more operations in parallel.

Choosing faster hardware modules may also allow for lower voltages [6]. There is a tradeoff here as in general faster hardware comes at the expense of a larger amount of switched capacitance.

Loop unrolling [1, 22], although not a transformation aiming at lower power in itself, can allow for a more effective use of pipelining and retiming. Loop unrolling may also increase the opportunity for variable sharing, reducing the amount of activity in the circuit [14].

Several transformation have been proposed for the reduction of switching activity. In [14] loop interchange is suggested as a way of increasing data locality, and thus, reducing the amount of data switching in a bus. Operand reordering may be used for the same purpose [1, 14].

The glitching component of the switching activity depends heav-

8. HIGH-LEVEL POWER ESTIMATION AND OPTIMIZATION   161

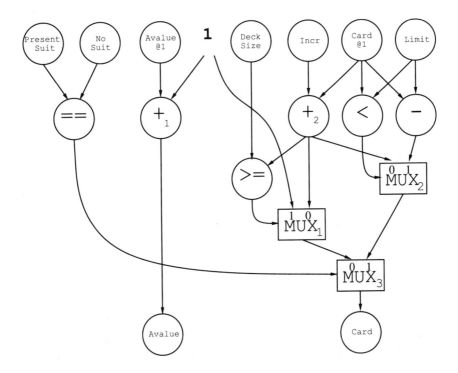

Figure 8.4: Control Data Flow Graph for the `dealer` circuit.

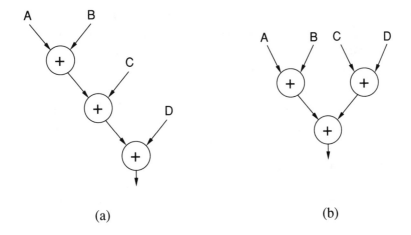

Figure 8.5: Chain vs. tree operations.

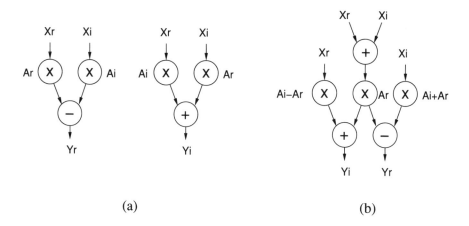

Figure 8.6: Trading a multiplication for an addition.

ily on the topology of the circuit. Balanced circuits like the tree topology of Figure 8.5(b) tend to have less glitching than the chain topology in Figure 8.5(a). Also, longer cascades of combinational blocks lead to a larger amount of glitching [1].

Power savings can be achieved by rearranging mathematical operations. Values that are used often can be computed in advance and stored instead recalculated. Further, this rearrangement can be used to tradeoff operations that require a larger amount switched capacitance for other more economic operations [1]. For example, assume we want to multiply some complex variable X with a complex constant A. By using the configuration of Figure 8.6(b) instead of the one in Figure 8.6(b), one multiplication, a much more power consuming operation, can be replaced by an addition.

More generally, multiplications with a constant can be replaced by some shift and add operations, at the expense of a more complex controller [1].

8.2.2 Scheduling Techniques

The scheduling process assigns a time slot for each operation in the CDFG. Throughput, cycle-time and hardware resource constraints make this a complex process. Even taking all precedence conditions in the CDFG into account, there is usually still some room for the order of operations. With this in mind, a scheduling algorithm that maximizes the potential for power

8. HIGH-LEVEL POWER ESTIMATION AND OPTIMIZATION

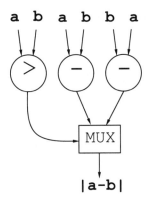

Figure 8.7: Control Data Flow Graph for $|a - b|$.

management in the resulting scheduled design is presented in [13].

In a typical design, the flow of data is determined at run time based on conditions derived from input values. As an example, say we need to compute $|a - b|$. One way to implement this is to do the comparison $a > b$ and if the result of this operation is true we compute $a - b$ otherwise we compute $b - a$. The Control Data Flow Graph (CDFG) for this simple example is shown in Figure 8.7. Assume that one control step is required for each of the three operations ($-$, $>$ and MUX).

The only precedence constraint for this example is that the multiplexor operation can only be scheduled after all other three operations. Existing scheduling algorithms use this flexibility to minimize the number of execution units needed and/or the number of control steps.

If we are allowed two control steps to compute $|a-b|$, then necessarily the operations $a > b$, $a - b$ and $b - a$ have to be executed in the first control step (we need two subtractors) and the multiplexor in the second control step as indicated in Figure 8.8.

If instead we are allowed three control steps, we can get by with one subtractor and schedule operations $a - b$ and $b - a$ in different control steps, one in the first control step and the other in the second. Operation $a > b$ can be scheduled in any of these two control steps and the multiplexor will be in the third control step, as shown in Figure 8.9.

In either case, *both $a - b$ and $b - a$ are computed although only the result of one of them is eventually used.* This is obviously wasteful in terms of power consumption.

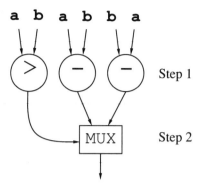

Figure 8.8: Schedule for $|a - b|$ using two control steps.

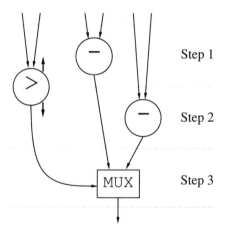

Figure 8.9: Schedule for $|a - b|$ using three control steps.

The scheduling algorithm of [13] attempts to assign operations involved in determining the data flow (in this case $a > b$) as early as possible in the initial control steps, thus indicating which computational units are needed to obtain the final result. Only those units that eventually get used are activated. The algorithm chooses a schedule only if the required throughput and hardware constraints are met. In other words, the algorithm explores any available slack to obtain a power manageable architecture.

For our example, and assuming we have available three control steps, the scheduling algorithm will assign $a > b$ to the first control step and $a - b$ and $b - a$ to the second. Depending on the result of $a > b$, only the inputs to one of $a - b$ and $b - a$ will be loaded, thus no switching activity will occur

8. HIGH-LEVEL POWER ESTIMATION AND OPTIMIZATION

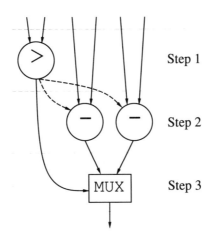

Figure 8.10: A power managed schedule for $|a - b|$ using three control steps.

in the subtractor whose result is not going to be used. This situation is shown in Figure 8.10, where the dashed arrows indicate that the execution of the '−' operations depends on the result of the comparator. Here we assumed we have two subtractors available. If that is not the case, we need to assign one subtract to the first control step and another to the second. The subtraction in the first control step will always be computed, but we can still disable the one in the second control step when it is not needed.

If only two control steps are allowed, there is no flexibility, the solution is unique (Figure 8.8), meaning that no power management is possible.

Based on the slack of the operations in the CDFG, obtained from the As Soon As Possible (ASAP) and As Late As Possible (ALAP) values, the scheduling algorithm of [13] determines which multiplexors in the circuit can be used for power management. Control edges are created in the CDFG starting at the node driving the control input of each selected multiplexor to the top nodes of the fanin of the 0,1-inputs of the same multiplexor. For the dealer example of Figure 8.4 (and for a given throughput), edges are created between node == and nodes < and − (this creates precedence conditions that ensure that also >=, MUX_1 and MUX_2 are scheduled after ==) and between node < and nodes − and $+_2$. The final CDFG for this example is shown in Figure 8.11.

The controller is now slightly more complicated since the execution or non-execution of some of operations depends on the result of previous

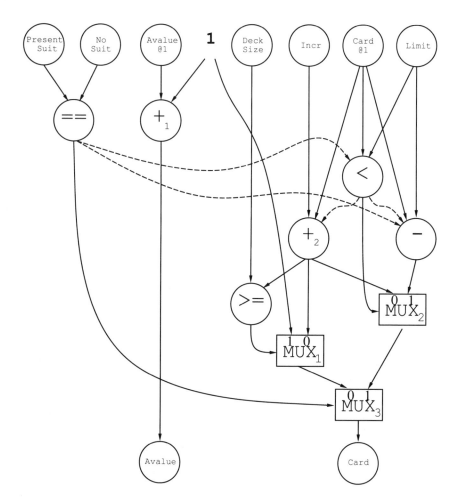

Figure 8.11: CDFG of `dealer` **with control edges for power management.**

operations. Returning to the `dealer` example, without power management the controller is shown in Figure 8.12(a). Notice that all operations are executed leading to wasteful power dissipation.

The controller with power management is shown in Figure 8.12(b). Depending on the result of the conditions associated with the multiplexors selected for power management, only signals that control operations that are going to be used are actually activated.

8. HIGH-LEVEL POWER ESTIMATION AND OPTIMIZATION

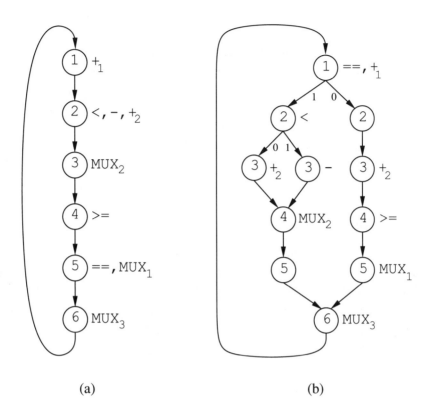

Figure 8.12: Controller (a)without and (b)with power management.

8.2.3 Allocation Techniques

During allocation, each operation is assigned to specific functional modules and operands to specific registers. The main objective is to minimize the number of required hardware modules and reduce the amount of interconnect. Recently techniques that also take power dissipation into account during allocation have been proposed.

The technique described in [20] targets minimizing the switching activity in registers and functional modules by a careful selection of which variables are assigned to the same register. Using functional simulation, statistics are obtained for each variable. From these statistics the correlation between variables can be computed. The algorithm selects variables with high correlation for the same register, keeping track on the complexity of the interconnect.

This way the number of transitions in the registers, and consequently in the functional modules, is minimized.

Similar approaches are described in [14] and [2].

An iterative method for simultaneous scheduling and allocation for low power is considered in [21]. Module selection is reevaluated after some temporary scheduling. An attempt is made to use slower modules (i.e., lower switched capacitance), even considering modules that take more that a clock period (multicycling). Further, the tradeoffs of resource sharing are weighed, the more sharing the less hardware is required at the expense of larger switching activity.

8.2.4 Optimizations at the Register-Transfer Level

Post behavioral synthesis optimization techniques for low power have also been proposed. The reduction of the amount of glitching present in a circuit described at the RT level is addressed in [18]. Several techniques are presented in order to reduce the glitching present in the control signals: using a modified multiplexor; multiplexor network restructuring; and gating control signals.

8.3 Conclusions

We have presented a survey of existing power estimation and optimization techniques that work at a high-level design description. These techniques are all very recent as this is a very new field of research. High-level tools rely on insight obtained from research at lower abstraction levels and their development began after lower level power tools matured.

At the register-transfer level, the circuit is described in terms of modules. The power dissipated by the circuit is computed from the power that each module consumes. At this level, a simple model of each module is used to make the estimation process very efficient. There is a penalty in accuracy which is acceptable as long as the relative power estimation between two designs is accurate enough to determine which is the most power dissipative. Approaches for the estimation of power dissipation for different RTL modules have been described.

The most effective optimization techniques for low power have been presented. Techniques that work at different steps in the synthesis process and which address different paramenters for low power have been surveyed. Among these, transformation techniques that target lower supply voltages described in Section 8.2.1 are among the most effective given the quadratic relationship between voltage and power. Significant power savings can also be achieved by power management methods, such as the one described in Section 8.2.2.

References

[1] A. Chandrakasan, M. Potkonjak, R. Mehra, J. Rabaey, and R. Broderson. Optimizing Power Using Transformations. *IEEE Transactions on Computer-Aided Design*, 14(1):12–31, January 1995.

[2] J. Chang and M. Pedram. Register Allocation and Binding for Low Power. In *Proceedings of the 32nd Design Automation Conference*, pages 29–35, June 1995.

[3] W. Donath. Placement and Average Interconnection Lengths of Computer Logic. *IEEE Transactions on Circuits and Systems*, 26(4):272–277, April 1979.

[4] M. Feuer. Connectivity of Random Logic. *IEEE Transactions on Computing*, 31(1):29–33, January 1982.

[5] D. Gajski, N. Dutt, A. Wu, and S. Lin. *High-Level Synthesis*. Kluwer Academic Publishers, 1992.

[6] L. Goodby, A. Orailoğlu, and P. Chau. Microarchitectural Synthesis of Performance-Constrained, Low Power VLSI Designs. In *Proceedings of the International Conference on Computer Design*, pages 323–330, October 1994.

[7] P. Hilfinger. A High-level Language and Silicon Compiler for Digital Signal Processing. In *Proceedings of the Custom Integrated Circuits Conference*, pages 213–216, May 1985.

[8] P. Landman and J. Rabaey. Architectural Power Analysis: the Dual Bit Type Method. *IEEE Transactions on VLSI Systems*, 3(2):173–187, June 1995.

[9] P. Landman and J. Rabaey. Activity-Sensitive Architectural Power Analysis. *IEEE Transactions on Computer-Aided Design*, 15(6):571–587, June 1996.

[10] D. Lidsky and J. Rabaey. Early Power Exploration - A World Wide Web Application. In *Proceedings of the 33rd Design Automation Conference*, pages 27–32, June 1996.

[11] D. Marculescu, R. Marculescu, and M. Pedram. Information Theoretic Measures for Power Analysis. *IEEE Transactions on Computer-Aided Design*, 15(6):599–610, June 1996.

[12] R. Mehra and J. Rabaey. Behavioral Level Power Estimation and Exploration. In *Proceedings of the International Symposium on Low Power Design*, pages 197–202, April 1994.

[13] J. Monteiro, S. Devadas, P. Ashar, and A. Mauskar. Scheduling Techniques to Enable Power Management. In *Proceedings of the 33rd Design Automation Conference*, pages 349–352, June 1996.

[14] E. Musoll and J. Cortadella. High-Level Synthesis Techniques for Reducing the Activity of Functional Units. In *Proceedings of the International Symposium on Low Power Design*, pages 99–104, April 1995.

[15] M. Nemani and F. Najm. Towards a High-Level Power Estimation Capability. *IEEE Transactions on Computer-Aided Design*, 15(6):588–598, June 1996.

[16] S. Powell and P. Chau. A Model for Estimating Power Dissipation in a Class of DSP VLSI Chips. *IEEE Transactions on Circuits and Systems*, 38(6):646–650, June 1991.

[17] S. Powell and P. Chau. Estimating Power Dissipation of VLSI Signal Processing Chips: the PFA Technique. In *VLSI Signal Processing IV*, pages 250–259, 1996.

[18] A. Raghunathan, S. Dey, and N. Jha. Glitch Analysis and Reduction in Register Transfer Level Power Optimization. In *Proceedings of the Design Automation Conference*, pages 331–336, June 1996.

[19] A. Raghunathan, S. Dey, and N. Jha. Register-Transfer Level Estimation Techniques for Switching Activity and Power Consumption. In *Proceedings of the International Conference on Computer-Aided Design*, November 1996. To appear.

[20] A. Raghunathan and N. Jha. Behavioral Synthesis for Low Power. In *Proceedings of the International Conference on Computer Design*, pages 318–322, October 1994.

[21] A. Raghunathan and N. Jha. An Iterative Improvement Algorithm for Low Power Data Path Synthesis. In *Proceedings of the International Conference on Computer-Aided Design*, pages 597–602, November 1995.

[22] M. Srivastava and M. Potkonjak. Power Optimization in Programmable Processors and ASIC Implementations of Linear Systems: Transformation-based Approach. In *Proceedings of the 33rd Design Automation Conference*, pages 343–348, June 1996.

[23] C. Svensson and D. Liu. A Power Estimation Tool and Prospects of Power Savings in CMOS VLSI Chips. In *Proceedings of the International Symposium on Low Power Design*, pages 171–176, April 1994.

Chapter 9

Conclusion

Rapid increases in chip complexity, increasingly faster clocks, and the proliferation of portable devices have combined to make power dissipation an important design parameter. The power dissipated by a digital system determines its heat dissipation as well as battery life. For some designs, power has become the most constringent constraint. Power reduction methods have been proposed at all levels – from system to device.

In this book we focused mainly on techniques at the logic level. At this abstraction level it is possible to use a simple but accurate model for power dissipation. The goal is to give the designer the ability to try different implementations of a design and compare them in terms of power consumption. For this purpose efficient power estimation tools are required.

9.1 Power Estimation at the Logic Level

The first part of this book was concerned with the problem of estimating the power dissipation of a logic circuit. In Chapter 2 the generally accepted model for power dissipation for static CMOS circuits described at the logic level was presented. It was shown that power dissipation is determined from the switching activity of the signals in the circuit, weighted by the capacitive load that each signal is driving.

Also in Chapter 2, we reviewed existing approaches for the switching activity estimation problem. These can be divided in two main categories: simulation-based and probabilistic techniques. The issues relative to each approach have been presented. Simulation-based techniques have the advantage that existing timing simulators can be used. The problem is then deciding how many input vectors are needed to obtain a desired accuracy level. For some circuits this may imply a long simulation run.

Probabilistic techniques can potentially be much more efficient, especially in the context of incremental modifications during synthesis. These approaches aim at propagating given primary input probabilities, static and/or transition, through the nodes in the circuit. Thus, in one pass the switching activity at each node can be computed. However, issues that are naturally handled in timing simulation arise for these probabilistic approaches, such as glitching and static and temporal correlation at primary inputs and internal nodes. The way a probabilistic approach deals with these problems determines its accuracy and run-time.

We have presented in Chapter 3 a probabilistic method that can handle these issues exactly. This approach is based on symbolic simulation. A Boolean condition for each node in the circuit making a transition at each time point is computed. This approach is efficient for circuits of small to moderate size (< 5000 gates). For larger circuits, the size of the symbolic network is too large for BDDs to be built.

In order to obtain accurate estimates for sequential circuits some other issues have to be taken into account. There exists a high degree of correlation between consecutive clock cycles. The values stored in the memory elements in the circuit at some clock cycle were generated in some previous clock cycle. Further, the probabilities at the output of these memory elements are determined by the functionality of the circuit and have to be calculated if accurate switching activity values are to be computed. Simulation-based techniques, though capable of taking the necessary correlation between clock cycles, have the problem of requiring a very large number of input vectors to be simulated before we can assume that steady state at the state lines has been achieved.

In Chapter 4 we presented an efficient way of computing the probabilities of state lines. This technique is applicable to circuits with an arbitrary number of memory elements. It requires the solution of a non-linear system of equations that can be solved using iterative methods. For the Picard-Peano method, we showed that, although only weak convergence proofs can be made, in practice the method works well and is faster than Newton-Raphson. For Newton-Raphson convergence conditions are met for most circuits. Previous techniques computed state probabilities as opposed to individual state lines

9. CONCLUSION

and were restricted to less than 20 registers. We pay some accuracy penalty by ignoring the correlation between the state lines; the experimental results show that the error introduced is less than 3% on average. Methods to improve this accuracy at the expense of computation time have also been presented.

Another problem we have addressed, also in Chapter 4, is the power estimation of a circuit given a particular input sequence. We described how a finite state machine can be built to model the input sequence. The methods for power estimation of sequential circuits can be applied to the cascade of this finite state machine and the original circuit. Other attempts to model the correlation of an input sequence involved computing correlation coefficients, typically between every pair of input signals. Besides the problem of having a large amount of information to specify to the power estimator, the accuracy can be low.

Future Work

The estimation of average switching activity and power dissipation in digital logic circuits is recognized as an important problem and no completely satisfactory solution has been developed. Hence a significant amount of research is being done on this problem.

The exact method we presented in Chapter 3, though efficient for small circuits, cannot be applied for large circuits and this an important limitation. It is generally accepted that approximation methods have to be used if circuits of significant size are to be handled.

Approximation schemes proposed for power estimation thus far lack some desirable properties. Most schemes are not based on an exact strategy, but based on heuristic rules that model correlation between internal signals in the circuit. While their runtime is typically polynomial, they are rarely parameterizable to improve accuracy at the expense of runtime, and are not calibrated against an exact strategy. Development of such approximation strategies is a subject of ongoing work.

9.2 Optimization Techniques at the Logic Level

Being a relatively new field and given its relevance for today's digital integrated circuits, optimization techniques for low power have been the subject of intense research in the last few years. The most representative work has been reviewed in Chapter 5.

At the logic level, power is directly related to the switched capacitance, i.e., switching activity of the signal weighted by the capacitance this signal is driving. We have presented in detail two different optimization methods. The first (Chapter 6) targets reduced glitching in the circuit by the use of retiming. The basic observation is that any glitching present at the input of a register is filtered by it. The registers are repositioned such that the reduction in switched capacitance is maximized. Up to 16% power reductions were obtained. The applicability of this technique is limited to pipelined circuits. The reason is that the operation of retiming in a cyclic sequential circuit changes the switching activity in the circuit globally thus it is very difficult to predict what the consequences of a particular move are.

We have described a more powerful optimization technique, termed precomputation, in Chapter 7. As stated above, the retiming technique is restricted to reducing the power dissipation due to glitching. Precomputation attempts the overall reduction of switching activity. A simple circuit is added to the original sequential circuit that tries to predict the circuit's outputs for the next clock cycle. When this is achieved, transitions at (all or part of) the inputs to the original circuit are prevented from propagating to the circuit by disabling the input registers. Significant power savings of up to 75% have been obtained and reported in the results section of Chapter 7.

Future Work

Research on optimization techniques for low power is under intense investigation; new approaches at all levels of abstraction will surely be proposed in the next few years.

At the logic level, we are lacking some scheme that can predict efficiently how the overall switching activity of a circuit is affected when some incremental change is done. This would be a very important method to guide re-synthesis tools for low power. Some work at this level has been proposed in [1].

A tool like the one described in the previous paragraph could allow for the extension of the retiming algorithm of Chapter 6 in order to find a global optimum for a k-pipeline, with $k \geq 1$ (instead of the iterative approach described in Chapter 6). It is also important to be able to handle cyclic sequential circuits such as finite state machines. For this purpose, some approximation has to be made regarding how the *inputs* to the registers change due to retiming (because of the feedback).

In Chapter 7 we described a few precomputation architectures. We have presented comprehensive results for the complete and subset input disabling sequential architectures. As to the multiplexor-based sequential architecture, some issues have still to be solved like on how many inputs to base the Shannon decomposition on and how to decide on which inputs to use.

More importantly, we believe that the combinational precomputation has potential that has not been completely explored. Better algorithms should be developed that decide on the optimum set of subcircuits to precompute.

9.3 Estimation and Optimization Techniques at the RT Level

Techniques at higher level of abstraction can have a higher impact on the power consumption of a circuit. It should be stressed however that the power savings obtained at different levels of abstraction are independent and add up to a less power consuming circuit. Further, understanding the impact that logic synthesis can have for low power design at the gate level can give a fundamental insight for the development of synthesis tools for low power at higher abstraction levels.

We made a survey of existing power estimation techniques at the register-transfer level in Section 8.1. At the RT level, the circuit is described in terms of modules and their interconnections. The approach for power estimation is to compute an estimation of the power dissipation for each module separately. Power dissipation models for different types of modules have been described in Section 8.1. These models are kept simple so that large designs can be efficiently handled. As a consequence, the accuracy of the power estimates will be lower than those obtained at the gate level. However, at the RT level the main objective is to be able to compare different design implementations and select the least power dissipative. Therefore, only a good relative power estimate is required.

Techniques for power optimization during behavioral synthesis have been surveyed in Section 8.2. Approaches that work at different steps in the behavioral synthesis process and that attack different parameters have been described. Given the quadratic relationship between the supply voltage and power, techniques that target lower supply voltages are very effective. The penalty for lower voltages is longer delays. This loss in throughput can be recovered, for instance, by pipelining.

We described in detail a scheduling technique that attempts to schedule the operations in an order such that controlling signals that decide the flow of data are computed first, thus indicating which operations are actually needed to compute the final result. The inputs to modules whose result would be discarded are disabled. More than 40% power savings can be achieved by this power-management-aware scheduling technique with little or no penalty in terms of hardware requirements. The limitations of this technique are that it is applicable only to designs where there is multiplexing of data.

References

[1] C. Lennard and A. Newton. An Estimation Technique to Guide Low Power Resynthesis Algorithms. In *Proceedings of the International Symposium on Low Power Design*, pages 227–232, April 1995.

Index

algebraic decision diagrams (ADD) 42
allocation 167
as late as possible (ALAP) 165
as soon as possible (ASAP) 165

binary decision diagram (BDD) 16, 26, 41, 103, 120, 138
boolean approximation method 19
branch prediction 73

Central Limit theorem 12
Chapman-Kolmogorov equations 40, 53
cofactor 17, 47, 116, 135, 147
completely-specified input sequence 66
contractive function 45
control data flow graph (CDFG) 159
control step 163
controllability don't-care set (SDC) 85
convergence 45, 47, 55, 62
correlation
    present state line 43, 52, 53
    primary input 65
    spatial 15, 65, 154
    temporal 13, 26, 38, 65, 154

delay model
    general 25

    unit 25
    zero 17, 24, 58
diagonally dominant matrix 51
directed acyclic graph (DAG) 99
disjoint sum-of-products 16
don't-care optimization 85
dual bit type method (DBT) 153
dynamic circuits 14
dynamic power 10

entropy 155
error function 12

factorization 86
finite state machine (FSM) 37, 89, 157

gated clock 91
glitching 16, 25, 97, 102, 154, 160, 168
graph covering 88
guarded evaluation 91

Hamming distance 89

incompletely-specified input sequence 67
inertial delay 29
information scaling factor 156
input-modeling FSM 66
interconnect sizing 84

invertible matrix 51

Jacobian matrix 48, 56

k-unrolled network 53, 63
kernel extraction 86

latch
    precomputation 142
    symbolic simulation 27
latency 160
leakage current power 10
legal retiming 100
linear programming (LP) 83
loop interchange 160
loop unrolling 160
low-density nodes 12
lower supply voltage 160

m-expanded network 52, 63
Markov process 40
maximum power dissipation 11
maximum-sized single-output circuits 144
Mealy machine 68, 91
Mean Value theorem 46, 50
Moore machine 67, 91
multicycling 168
multiple cycle precomputation 126

Newton-Raphson method 47, 56, 58, 62
next state logic 38
norm
    ∞-norm 46
    1-norm 49

observability don't-care set (ODC) 91, 116, 133

operand reordering 160

Parker-McCluskey method 19
path balancing 84
Picard-Peano method 45, 55, 62
pipelined circuit
    power estimation 36, 58
pipeline 35, 99
pipelining 159
power dissipation model 10
precedence conditions 159
precomputation architecture
    combinational precomputation 141
    complete input disabling 132
    datapath examples 125
    logic duplication 122
    multiplexor-based 147
    output selection 120
    subset input disabling 113
precomputation logic 116
predictor functions 113
present state line correlation 43, 52, 53
present state line probabilities 39, 43, 61
primary input correlation 65
probabilistic estimation
    issues 13
    techniques 16

random logic simulation 11, 26
reconvergent fanout 15
redundant state lines 54
reliability 4, 12
resource sharing 160, 168
retiming 97, 160
RTL estimation

INDEX

    controller 157
    functional modules 152
    interconnect 158
satisfiability don't-care set (SDC) 85
scheduling 162
sensitivity 102
sequential circuit
    acyclic 35
    cyclic 37
sequential power estimation
    approximate 42, 58, 73
    exact 41, 58
Shannon circuits 84
Shannon expansion 47, 147
short-circuit power 10, 83
shut-down 90, 111
signal encoding 90
Silage 159
simulation-based estimation 11, 42
simulation
    cycle-based 154
    logic 11
    switch-level 9
    symbolic 24
SIS 32, 77, 129, 140
slack 164
software power analysis 66, 69
SPICE 9
spurious transitions 16, 25, 97, 102, 154, 160, 168
state encoding 89
state probabilities 38, 39
state transition graph (STG) 38, 54, 58, 66, 89, 157
static circuits 14
static probability 13

support of a function 122
switch-level simulator 9
switched capacitance 82, 153, 160
switching activity estimation 11
symbolic simulation 24

technology mapping 87
test generator (ATG) 3
throughput 159
transistor sizing 82
transition density 10, 17
transition probability 13
transition waveform 18
transmission gates
    symbolic simulation 27

universal quantification 117

wave-pipelining 84

yield 4